CHRISTOPHER HARRIS
SCOTT A. HARVEY

Hazardous Chemicals and the Right to Know

AN UPDATED GUIDE TO COMPLIANCE WITH SARA TITLE III

McGraw-Hill, Inc.

New York San Francisco Washington, D.C. Auckland Bogotá
Caracas Lisbon London Madrid Mexico City Milan
Montreal New Delhi San Juan Singapore
Sydney Tokyo Toronto

Executive Enterprises Publications Co., Inc.

New York

Library of Congress Catalog Card Number 93-40613

1 2 3 4 5 6 7 8 9 0 KGP/KGP 9 9 8 7 6 5 4 3

ISBN 0-07-026906-8

The sponsoring editor for this book was Gail F. Nalven and the production supervisor was Donald F. Schmidt.

Printed and bound by Arcata Graphics/Kingsport.

This book is printed on acid-free paper.

Previously published as *SARA Title III: Hazardous Chemicals and the Right to Know.*

Hazardous Chemicals and the Right to Know

Other McGraw-Hill Environmental Engineering Books of Interest

AMERICAN WATER WORKS ASSOCIATION • *Water Quality and Treatment*
BAKER • *Bioremediation*
CHOPEY • *Environmental Engineering for the Chemical Process Industries*
CORBITT • *Standard Handbook of Environmental Engineering*
FREEMAN • *Hazardous Waste Minimization*
FREEMAN • *Standard Handbook of Hazardous Waste Treatment and Disposal*
JAIN • *Environmental Impact Assessment*
LEVIN, GEALT • *Biotreatment of Industrial and Hazardous Waste*
MAJUMDAR • *Regulatory Requirements of Hazardous Materials*
McKENNA, CUNNEO • *Pesticide Regulation Handbook*
NANNEY • *Environmental Risks in Real Estate Transactions*
WALDO, HINES • *Chemical Hazard Communication Guidebook*

Table of Contents

Chapter Two

Section 304: Emergency Release Notification 9

Chapter Three

Chapter Four

Chapter Six

Appendices

About the Authors

Christopher Harris, a partner in McCutchen, Doyle, Brown & Enersen's Washington, DC, office, has extensive experience in legislative and regulatory projects, as well as in appellate litigation. During the 98th Congress, he served as counsel to the House Energy and Commerce Subcommittee, with jurisdiction over the Resource Conservation and Recovery Act (RCRA) and Superfund. In that capacity, he was the staff attorney responsible for drafting and managing the 1984 amendments to RCRA. He was with the Justice Department's Land and Natural Resources Division from 1979 to 1983 and worked primarily on RCRA and Superfund liability issues. In 1980, Mr. Harris received the Attorney General's Special Achievement Award for his efforts in drafting the criminal provisions in the 1980 amendments to RCRA. Previously, he served as chief counsel to a Senate Judiciary Subcommittee, and in the Office of General Counsel of the U.S. General Accounting Office. Mr. Harris has written books on a variety of topics, including criminal prosecutions for environmental violations, reporting of releases of hazardous substances, used oil recycling and hazardous waste legislation. Mr. Harris serves as General Counsel to the National Oil Recyclers Association. A graduate of Yale University, he served as president of the Yale Political Union. He holds a master of science degree from the London School of Economics and a law degree from Boston College Law School. Mr. Harris is admitted to practice in Massachusetts and the District of Columbia and to appear before the U.S. Supreme Court.

Scott Alan Harvey is an associate at Miles & Stockbridge in the Washington, DC, office. He obtained his B.A. degree with honors in Government and Foreign Affairs from the University of Virginia in 1988 and his J.D. from the University of Virginia in 1991. He is a member of the Pennsylvania and Virginia bars. Mr. Harvey counsels manufacturing and construction clients on Resource Conservation and Recovery Act (RCRA), Toxic Substances Control Act (TSCA), Comprehensive Environmental Response, Compensation, and Liability Act (CERCLA), and Emergency Planning and Community Right-to-Know Act (EPCRA) matters.

Introduction

In the wake of the tragedy in Bhopal, India—a release of methyl isocynate, a pesticide intermediate, from a Union Carbide plant, which resulted in the deaths of more than 2,000 people and serious injuries to many more—Congress became acutely aware of the environmental and safety hazards posed by the storage and handling of toxic chemicals. According to Senator Robert Stafford of Vermont, the "Bhopal disaster focused public attention on the fact that extremely dangerous chemicals are present at chemical manufacturing plants and other facilities in communities all across America." In Congress's view, the existence of this danger conferred on citizens a "right to know about these chemicals—what they are, where they are, and how much of them is present."

The legislative response to the Bhopal incident, designed to prevent similar tragedies in the United States, is a comprehensive system of reporting, data gathering, and information sharing that has generated thousands of tons of paper and redirected the activities of thousands of individuals in both the public and private sectors. The debate in Congress on the merits of a federal community right-to-know program, although acknowledging the enormity of the task and risks to confidential business information, focused relentlessly on the danger posed by toxic chemicals to an uninformed public. The remarks of Representative Bob Edgar of Pennsylvania exemplified the mood on Capitol Hill.

> This is a new Federal initiative, and I recognize the desire of some of my colleagues to move ahead cautiously to ensure the burdens imposed on industry are not excessive. Frankly, my concerns rest with the families that live in the shadow of these chemical and manufacturing plants.

The Senate sponsors of the legislation were equally adamant. According to Senator Frank Lautenberg of New Jersey,

> The right to know means public information about what hazardous substances are being stored and released into the environment. It means that our citizens and our emergency response personnel will be safer from chemical releases. It means this Nation will not tolerate Bhopal- or Chernobyl-type tragedies.

Concerns about the legislation were expressed, not as doubts about the fundamental wisdom of enacting a comprehensive new law, but rather about the potential for excessive regulatory burdens. Representative Norman F. Lent of New York, a key sponsor of the 1986 Superfund amendments, stated:

> Although I support the concept of a community's right to know of the risks its businesses pose, I am concerned that the legislation, if misinterpreted, could result in any useful information being buried in an avalanche of unnecessary paperwork.

A few other members of Congress, such as Representative John Hammerschmidt of Arkansas, worried not just about the "tremendous burden on industry" but also about the "unmanageable burden on state and local government."

> There are plenty of requirements, but no funds to help State and local governments comply with the law. Consequently, there is the potential for massive noncompliance; and when companies do comply with the law, State and local agencies will be inundated with paperwork, thereby rendering the whole exercise useless.

Whether the implementation of Congress's plan to provide communities with a "right to know" is unfolding according to its expectations—or even whether the plan was a realistic one in the first place—remains to be seen. What is clear is that this law and its implementing regulations are so complex that the success of this program will depend, in large measure, on the diligence, ingenuity, and cooperation of all participants. This is particularly apparent when one considers the failure of the federal government to provide adequate funding to state and local governments to defray the cost of implementing Congress's mandate.

The Emergency Planning and Community Right-to-Know Act of 1986 (EPCRA, or SARA Title III) can be seen as a logical expansion of the worker right-to-know programs that are mandated under the Occupational Safety and Health Act of 1970 (OSHA). In enacting OSHA, Congress directed that the Secretary of Labor "shall prescribe the use of labels or other appropriate forms of warning as are necessary to insure that employees are apprised of all hazards to which they are exposed, relevant symptoms and appropriate medical emergency treatment and proper conditions and precautions of safe use or exposure." Thirteen years after OSHA's enactment, on November 25, 1983, the Department of Labor's Occupational Safety and Health Administration promulgated regulations requiring employers to provide information concerning hazardous chemicals to their employees. These regulations, which became known as the Hazard Communication Standard, require employers to place labels on containers of hazardous substances used in the workplace, prepare material safety data sheets (MSDSs), and provide training to employees regarding safe handling of chemicals.

Although implementation of the Hazard Communication Standard is no small task given the fact that workers in the United States are exposed to an estimated 575,000 hazardous chemicals, Congress—seemingly oblivious to the regulatory burden—decided that the right to information on these chemicals should be extended to every U.S. citizen. The result, which is much more than a mere extension of the Hazard Communication Standard, is one of the most intricate paper chases that industry has had to undertake. Thus, although the new law builds on the foundation laid by OSHA, data retrieval and reporting have created an elaborate regulatory structure generating an avalanche of paper. Nevertheless, EPCRA and its implementing regulations (in combination with the Pollution Prevention Act of 1990) have had the effect of focusing the attention of corporate management on the vast quantity of hazardous chemicals used and released by American industry each year. This has led to genuine progress in waste recycling and source reduction.

The purpose of this book is to review and explain the elements of federal right-to-know legislation, focusing on industry's regulatory obligations, including reporting requirements such as the submission of chemical inventories and quantification of toxic chemical releases. In addition, this book includes various fact-specific inserts, entitled "EPA Guidance," which provide the reader with insight concerning EPA's policies regarding EPCRA. These inserts feature information provided by EPA's EPCRA hotline staff in Washington, DC. Although the information in the inserts is not legally binding on EPA, it provides the reader with useful information on how EPA will likely construe ambiguities in EPCRA's reporting requirements and apply EPCRA's provisions to unusual fact situations. Finally, to assist the reader in traversing this regulatory labyrinth, various tables, flow charts, and reporting forms have been included as appendices.

General Overview of Reporting Requirements

As the length of this book attests, the Emergency Planning and Community Right-to-Know Act gives rise to a variety of complicated regulatory issues that affect property owners and businesses that are involved with hazardous substances, chemicals, and wastes. Providing appropriate information to the right authorities in a timely manner can pose significant burdens on members of the regulated community. Indeed, in many instances, it is difficult to determine whether particular individuals or business entities are subject to EPCRA in the first place.

When negotiating one's way through EPCRA's regulatory maze, it is easy to lose sight of the Act's principal objectives. Although the application of EPCRA to particular business activities can create complicated legal issues, the fundamental requirements of the Act are quite simple. If one keeps these fundamental requirements in mind when reading the more detailed discussions in the book, making decisions on how best to comply with EPCRA in particular situations should be a less difficult endeavor.

EPCRA imposes five basic reporting requirements on members of the regulated community:

1. Section 302 creates emergency planning requirements.

2. Section 304 imposes emergency release reporting obligations.

3. Section 311 requires the submission of material safety data sheets.

4. Section 312 requires the submission of chemical inventory forms.

5. Section 313 requires the submission of toxic chemical release reporting forms.

The obligation of a facility owner or operator to comply with these requirements depends on the type of facility, the amount and type of hazardous material present at a facility or released from a facility, and the activities a facility engages in that involve hazardous material.

A facility owner or operator is subject to the emergency planning requirements of Section 302 if an extremely hazardous substance (EHS) becomes present at the facility in excess of a threshold planning quantity (TPQ). EHSs and TPQs appear on a list published by the Environmental Protection Agency (EPA). EPCRA requires the owner or operator to determine whether a TPQ is exceeded. If a TPQ is exceeded, the owner or operator must notify the appropriate state emergency response commission (SERC), appoint an emergency coordinator for the facility, and provide information about the facility to a local emergency planning committee (LEPC).

A facility owner or operator is subject to the emergency release reporting requirements of Section 304 if a reportable quantity (RQ) or more of an EHS or a Comprehensive Environmental Response, Compensation, and Liability Act (CERCLA) hazardous substance is released from a facility. EPA publishes lists of EHSs, CERCLA hazardous substances, and their RQs. The owner or operator must report releases to the appropriate SERC and LEPC immediately, and must provide a written follow-up report as soon as practicable after the release. If a CERCLA hazardous substance is released, the owner or operator must also report the release to the National Response Center.

A facility owner or operator becomes subject to the MSDS and hazardous chemical inventory reporting requirements of Sections 311 and 312 if OSHA requires the facility to develop or maintain an MSDS for a hazardous chemical *and* the hazardous chemical becomes present at the facility in excess of a threshold reporting quantity (TRQ). Hazardous chemicals include all chemicals that present physical or health hazards. The TRQ for most hazardous chemicals is 10,000 pounds. The TRQ for hazardous chemicals that are also EHSs is 500 pounds or the TPQ listed by EPA pursuant to Section 302, whichever is lower. When a TRQ for a hazardous chemical is exceeded, the facility owner or operator must submit an MSDS and a hazardous chemical inventory form to the SERC, LEPC, and local fire department with jurisdiction over the facility. MSDSs provide these agencies with information concerning the physical and health hazards associated with hazardous chemicals that are present at a facility. Inventory forms provide information concerning the quantity and location of hazardous chemicals at a facility.

Owners or operators of manufacturing facilities that manufacture or process 25,000 pounds, or otherwise use 10,000 pounds, of a toxic chemical during a calendar year are subject to the toxic chemical release reporting requirements of Section 313. EPA publishes a list of toxic chemicals that trigger reporting obligations. The owner or operator must complete an EPA Form R for each toxic chemical that is manufactured, processed or otherwise used at the facility in excess of a reporting threshold. Form R requires owners and operators to use available data to estimate the quantity of toxic chemicals released into the environment or transported off-site during a calendar year. Form R also requires owners and operators to provide information regarding the quantity of toxic chemicals recycled on-site, the quantity sent off-site for recycling, and the pollution reduction techniques, if any, that a facility uses to reduce releases of toxic chemicals. Forms must be filed with EPA and designated state agencies.

The following chapters provide detailed information concerning the five reporting requirements outlined above, including practical examples and information on how EPA applies EPCRA to specific business situations. The enforcement of EPCRA's reporting provisions under Sections 325 and 326, and procedures members of the regulated community can employ to withhold information pursuant to the trade secret provisions of Section 322 is also explained. When applying this information to develop compliance strategies for particular businesses, it is important to keep in mind that EPCRA's basic provisions are designed to encourage businesses to generate and make available information regarding hazardous material. In order to minimize the risks of prolonged and expensive enforcement actions, facility owners and operators should, whenever possible, resolve ambiguities in the reporting provisions in favor of disclosing, as opposed to withholding, information.

Table of Abbreviations

CERCLA	*Comprehensive Environmental Response, Compensation, and Liability Act (Superfund)*
EHS	*Extremely Hazardous Substance*
EPA	*United States Environmental Protection Agency*
EPCRA	*Emergency Planning and Community Right-to-Know Act (SARA Title III)*
FDA	*Food and Drug Administration*
FIFRA	*Federal Insecticide, Fungicide, and Rodenticide Act*
LEPC	*Local Emergency Planning Committee*
MSDS	*Material Safety Data Sheet*
NPDES	*National Pollutant Discharge Elimination System*
OSHA	*Occupational Safety and Health Act*
RCRA	*Resource Conservation and Recovery Act*
RQ	*Reportable Quantity*
SERC	*State Emergency Response Commission*
TCLP	*Toxic Characteristic Leaching Procedure*
TPQ	*Threshold Planning Quantity*
TRQ	*Threshold Reporting Quantity*
TSCA	*Toxic Substances Control Act*

Chapter One

EPCRA Section 302: Emergency Planning Requirements

One of the primary objectives of the Emergency Planning and Community Right-to-Know Act is to prepare local organizations to respond to accidental releases of hazardous substances and increase public awareness of the use of hazardous substances in local communities. EPCRA requires states to set up organizations to achieve these objectives, and requires members of the regulated community to provide them with sufficient information to accomplish their tasks.

State and Local Emergency Planning Organizations

State Emergency Response Commissions

As mandated by Section 301 of EPCRA, the governor of each state has established a state emergency response commission (SERC).[1] The commissions are composed of individuals with special technical expertise and experience in responding to emergencies involving hazardous substances.[2] They are responsible for overseeing the emergency planning process. EPCRA requires SERCs to divide states into emergency planning districts and appoint local emergency planning committees (LEPCs) to develop emergency response plans for each district.[3] The SERCs review emergency response plans[4] and work with LEPCs to develop procedures for providing information to the public regarding the production, storage, and use of hazardous substances in local communities.[5]

---

1. A list of SERCs, including contacts, addresses and telephone numbers is provided at **Appendix A**.
2. See Section 301(a), 42 U.S.C. §11001(a) (requiring governors to appoint persons with experience in the emergency response field to the extent practicable).
3. Section 301(b)-(c), 42 U.S.C. §11001(b)-(c). EPA does not publish a comprehensive list of LEPC contacts. Members of the regulated community should contact the SERC with jurisdiction over their facility to obtain a list of LEPC contacts.
4. Section 303(e), 42 U.S.C. §11003(e).
5. EPCRA requires emergency response plans, material safety data sheets, hazardous chemical lists, hazardous chemical inventory forms, toxic release inventory forms, and follow-up emergency release reports to be made available for public inspection. Section 324(a), 42 U.S.C. §11044(a). SERCs and LEPCs are jointly responsible for setting up procedures to provide the public with such information. See Section 301(a), (c), 42 U.S.C. §§11001(a),(c).

Local Emergency Planning Committees

LEPCs have jurisdiction over the emergency planning process for each planning district.[6] They assume primary responsibility for collecting information from regulated facilities and developing emergency response plans.[7] LEPC members typically include elected officials; members of law enforcement, firefighting, hospital, transportation, and civil defense organizations; journalists; and owners and operators of industrial facilities.[8]

Each LEPC must develop an emergency response plan for the planning district under its jurisdiction. LEPCs review the plans at least once a year and revise them if necessary.[9] At a minimum, the plans must include:

1. Identification of facilities subject to [EPCRA] that are within the emergency planning district, identification of routes likely to be used for the transportation of [EHSs] and identification of additional facilities contributing or subjected to additional risk due to their proximity to [covered facilities], such as hospitals or natural gas facilities;
2. Methods and procedures to be followed by facility owners and operators and local emergency personnel to respond to any release of such substances;
3. Designation of a community emergency coordinator and facility emergency coordinators, who shall make determinations necessary to implement the plan;
4. Procedures for providing reliable, effective, and timely notification by the facility emergency coordinators and the community emergency coordinator to persons designated in the emergency plan, and to the public, that a release has occurred (consistent with the emergency notification requirements of Section 304 of EPCRA);
5. Methods for determining the occurrence of a release, and the area or population likely to be affected by such release;
6. A description of emergency equipment and facilities in the community and at each [covered facility] in the community and an identification of the persons responsible for such equipment and facilities;
7. Evacuation plans, including provisions for a precautionary evacuation and alternative traffic routes;
8. Training programs, including schedules for training of local emergency response and medical personnel; and

---

6. See Section 301(c), 42 U.S.C. §11001(c).
7. See Section 303(d), 42 U.S.C. §11003(d)(3) (authorizing LEPCs to collect information that is necessary to develop emergency response plans); Section 303(a), 42 U.S.C. §11003(a) (empowering LEPCs to draft and review emergency response plans).
8. See Section 301(c), 42 U.S.C. §11001(c).
9. Section 303(a), 42 U.S.C. §11003(a).

9. Methods and schedules for exercising the emergency plan.[10]

Facilities Subject to EPCRA Planning Requirements

Facilities with Extremely Hazardous Substances On-Site in Excess of Threshold Planning Quantities

EPCRA requires owners or operators of facilities[11] to notify the appropriate SERC[12] when they become subject to EPCRA threshold planning requirements. Facilities become subject to EPCRA threshold planning requirements when an extremely hazardous substance (EHS) is present at the facility in excess of a threshold planning quantity (TPQ).[13]

To determine whether a TPQ has been exceeded, the owner or operator must determine the total quantity of an EHS that is present at the facility—regardless of location, number of containers, or method of storage.[14] EPCRA does not require owners or operators to test mixtures or ask manufacturers to provide them with information regarding EHSs in mixtures. However, when calculating TPQs, they must take into account the quantity of an EHS present in mixtures or solutions that they know, or reasonably should know, contain an EHS.[15]

EXAMPLE: A facility has 500 pounds of pure EHS X on-site. The TPQ for EHS X is 1,000 pounds. A supplier delivers 1500 pounds of a mixture to the facility. Along with the shipment, the supplier provides a material safety data sheet and a supplier notification form indicating the mixture is 50 percent EHS X by weight. Given the information accompanying the shipment, the owner or operator of the facility reasonably should know how much EHS X is present at the facility (1,250 lbs.),

---

10. Section 303(c)(1)-(9), 42 U.S.C. §11003(c)(1-9). A detailed guide to emergency planning is provided at **Appendix B**. The Minnesota Department of Public Safety developed this guide to aid LEPCs in their planning efforts. Reference to this document may help owners and operators of covered facilities anticipate the type of information LEPCs will require from them in order to develop emergency response plans.

11. For threshold planning purposes, the term "facility" includes "all buildings, equipment, structures, and other stationary items that are located on a single site or on contiguous or adjacent sites that are owned and operated by the same person (or by any person that controls, is controlled by, or under common control with, such person). "Facility" shall include man-made structures in which chemicals are purposefully placed or removed through human means such that it functions as a containment structure for human use. For purposes of emergency release notification, the term includes motor vehicles, rolling stock and aircraft. 40 C.F.R. §355.20(a).

12. The "appropriate SERC" is the one that has jurisdiction over the geographical location of the facility. Section 302(c), 42 U.S.C. §11002(c).

13. See Section 302(c), 42 U.S.C. §11002(c) outlining notification requirements. See also Section 302(b)(1), 42 U.S.C. §11002(b)(1). EPA publishes a list of EHSs and corresponding TPQs at 40 C.F.R. §355, **Appendix A**. A comprehensive list of EHSs, including TPQs is provided at **Appendix C**.

14. 52 Fed. Reg. 13378, 13380 (April 22, 1987).

15. 52 Fed Reg. at 13385.

and should notify the appropriate SERC that the facility is subject to threshold planning requirements.

EPA Guidance

Calculating TPQs When the Precise Concentration of EHS in a Mixture Is Unknown

If the owner/operator of a facility knows only the range of concentration of an EHS that is present in a mixture, the upper-bound concentration must be used to calculate the quantity of an EHS that is present at a facility.

EXAMPLE: 50,000 pounds of a mixture that contains EHS X is present at a facility. The MSDS for the mixture indicates that the mixture is between 35 percent and 50 percent EHS X by weight. For reporting purposes, 25,000 pounds of EHS X is present at the facility as part of the mixture.

This rule applies to all situations under Sections 302, 304, and 311/312, in which EPCRA requires owners and operators to determine whether a threshold planning quantity, reportable quantity, or threshold reporting quantity has been exceeded.

EPCRA HOTLINE February 1990

Owners or operators of so-called covered facilities must notify the appropriate SERC within sixty days of the date that their facilities become subject to EPCRA threshold planning requirements.[16] A facility can become subject to threshold planning requirements:

1. When it accumulates an EHS on-site in excess of a threshold planning quantity.

EXAMPLE: On January 10, 1992, a facility has 1,000 pounds of EHS Y present on-site. The TPQ for EHS Y is 5,000 pounds. On January 15, 1992, the facility receives a ship-

16. Section 302(c), 42 U.S.C. §11002(c); 40 C.F.R. § 355.30(b).

ment of 9,000 additional pounds of Substance Y, which it stores on-site. The owner or operator of the facility must notify the SERC within sixty days of January 15.

2. When EPA revises its list of EHSs and TPQs in a manner that creates a new reporting obligation.

EXAMPLE: For the past several years, a facility has stored, and continues to store, at least 1,000 pounds of Substance Z on-site. On January 15, 1992, EPA classifies Substance Z as an EHS for the first time, and establishes a TPQ of 500 pounds. The owner or operator of the facility must notify the SERC within sixty days of January 15.

Form of Notice

Notification to the SERC must be in writing and must specify the name and street address of the facility.[17]

Multiple Notifications

Owners or operators must notify SERCs each time their facilities become subject to threshold planning requirements for an additional EHS.[18] They should also notify the LEPC with jurisdiction over their facility. EPCRA requires owners or operators to provide LEPCs with notice of all changes that occur at covered facilities that are relevant to the emergency planning process.[19] Presumably, the accumulation of an additional EHS on-site in excess of a TPQ would constitute such a change.

EXAMPLE: On February 1, 1989, a facility accumulated EHS A on-site in excess of its TPQ. Within sixty days, the owner notified the SERC that the facility was subject to EPCRA for the first time. On January 15, 1992, the facility accumulates EHS B on-site in excess of its TPQ. The owner of the facility must notify the SERC and the LEPC that substance B is present at the facility in excess of a TPQ.

De Minimis Concentration Exception

EHSs are not considered to be present at facilities when they constitute less than 1 percent of a mixture or solution by weight.[20]

---

17. 52 Fed. Reg. 13379.
18. Bureau of National Affairs, *Right to Know Planning Guide* §541:2001 (Oct. 1992).
19. See Section 302(d)(2), 42 U.S.C. §11003(d)12); 40 C.F.R. §355.30(d).
20. 40 C.F.R. §355.30(a), (e).

EXAMPLE: A facility stores one hundred thousand pounds of a liquid mixture that contains 99.5 percent nonhazardous material and .5 percent toluene 2, 4-diisocyanate, an EHS with a TPQ of 500 pounds. The facility is not subject to EPCRA threshold planning requirements. Even though toluene 2, 4-diisocyanate is present on-site in an amount equal to its TPQ, its concentration is too low to subject the facility to EPCRA planning requirements.

Special TPQs for EHSs that Are Solids

EPA establishes different TPQs for solids depending on the form they take when present at facilities. For example, many EHSs have a TPQ of "500/10,000 pounds."[21] Owners and operators must use the lower TPQ for solids that are present at facilities in powdered form with a particle size less than 100 microns, in solution or molten form, or in a form that meets the criteria for a National Fire Protection Association rating of 2, 3, or 4 for reactivity.[22]

Specially Designated Facilities

The governors of each state or the SERCs may declare certain facilities to be subject to EPCRA threshold planning requirements even though they do not have an EHS on-site in excess of a TPQ. The governor or SERC must provide interested parties with notice and an opportunity to comment on proposed designations.[23] After a facility has been so designated, it must comply with the same requirements that are applicable to other covered facilities.

▼ Threshold Planning Requirements

Owners or operators of covered facilities must comply with three simple threshold planning requirements. They must:

1. Designate a facility representative to serve as an "emergency response coordinator" and work with the LEPC to develop and implement an emergency response plan;[24]
2. Provide information that is necessary for the development and implementation of the local emergency response plan upon request to the LEPC;[25] and

---

21. The special TPQ listings for EHSs are set forth in **Appendix C**.
22. 40 C.F.R. §355.30(e)(2)(i).
23. Section 302(b)(2), 42 U.S.C. §11002(b)(2); 40 C.F.R. §355.30(a).
24. Section 303(d)(1), 42 U.S.C. §11003(d)(1); 40 C.F.R. §355.40(c).
25. Section 303(d)(3), 42 U.S.C. §11003(d)(3); 40 C.F.R. §355.40(d)(2). A sample LEPC information request is provided at **Appendix D**.

3. Inform the LEPC of any changes that occur at the facility that may be relevant to emergency planning.[26]

▼ Joint and Several Liability

Property owners should note that neither Section 302 of EPCRA nor its implementing regulations contain provisions exempting landlords from liability for their tenants' failure to comply with threshold notification and emergency planning requirements.[27] EPA may hold the owner *or* operator of a facility liable for reporting violations *regardless* of who is actually in day-to-day control of a facility. Property owners should keep this potential liability in mind when they lease land to industrial tenants.

EPA Guidance

EPCRA Does Not Apply to Operators of Federal Facilities: Persons Who Lease Property to Federal Facilities May Be Solely Responsible for EPCRA Compliance

Persons who lease property to the federal government should be aware that they are responsible for notifying appropriate SERCs and complying with EPCRA planning requirements, even though they lack any business interest or day-to-day control over facilities on their property.

EXAMPLE: The Department of the Army leases property from a private individual and begins to operate a laboratory on the property. The Army totally controls the laboratory. The owner of the land has only a real estate interest in the property and does not control the activities at the laboratory. The owner of the land has no employees

continued

---

26. Section 303(d)(2), 42 U.S.C. §11003(d)(2); 40 C.F.R. §355.40(d). Neither EPCRA, its implementing regulations, nor EPA policy documents specify what type of information may be relevant to the emergency planning process. Covered facilities must determine what is relevant based on their working relationships with LEPC members and officials.

27. In contrast, EPCRA specifically exempts owners from liability for toxic chemical release reporting violations under Section 313, providing their interest in a facility is limited to an interest in real estate. 40 C.F.R. §372.38(e).

continued

at the property. Because EPCRA's definition of "person" does not include federal facilities, the Department of the Army need not report. Because EPCRA requires owners *or* operators to comply with reporting requirements, the owner in this case must satisfy all EPCRA planning requirements that the Department of the Army does not voluntarily satisfy.

EPCRA HOTLINE February 1990

Chapter Two

Section 304: Emergency Release Notification

Another of EPCRA's primary purposes is to ensure that communities receive adequate warning in the event of a release of a hazardous substance. Early warning allows local emergency response personnel to implement emergency response plans more effectively, thereby protecting human health and preserving the environment. EPCRA requires owners or operators of facilities that produce, store, and use hazardous chemicals[1] to report releases of EHSs and CERCLA hazardous substances that in a twenty-four-hour period exceed reportable quantities.[2] Reports must be made to the SERCs and emergency coordinators for the LEPCs of every planning district likely to be affected by a release.[3] In addition, persons in charge of facilities must report releases of CERCLA hazardous substances to the National Response Center.[4]

Substances that Trigger Emergency Reporting Obligations

Extremely Hazardous Substances

EPCRA authorizes EPA to designate a substance as extremely hazardous and assign it an RQ based on its toxicity and its potential to become dispersed into the environment in the event

---

1. The term "hazardous chemicals" includes all hazardous chemicals defined in 29 C.F.R. §1910.1200, except: (1) any food, food additive, drug, or cosmetic regulated by the Food and Drug Administration; (2) any substance present as a solid in any manufactured item to the extent exposure to the substance does not occur under normal conditions of use; (3) any substance to the extent it is used for personal, family, or household purposes, or is present in the same form and concentration as a product packaged for distribution and use by the general public; (4) any substance to the extent it is used in a research laboratory or a hospital or other medical facility under the direct supervision of a technically qualified individual; and (5) any substance to the extent it is used in routine agricultural operations or is a fertilizer held for sale by a retailer to the ultimate customer.
2. Section 304(a), 42 U.S.C. §11004(a); 40 C.F.R. §355.40(a). Neither EPCRA nor its implementing regulations expressly state that an RQ must be released within twenty-four hours to trigger reporting obligations. However, EPA has adopted the twenty-four hour period as a matter of policy. See 52 Fed. Reg. 13386 (EPCRA's emergency release reporting requirement is similar to CERCLA's).
3. Section 304(b)(1), 42 U.S.C. §11004(b)(1); 40 C.F.R. §355.30(b). A "community emergency coordinator" is a state official appointed by the LEPC.
4. Section 304(a)(1), (3), 42 U.S.C. §11004(a)(1), (3); 40 C.F.R. §355.40, note to paragraph (a).

of an accidental release.[5] EPA publishes a list of EHSs and RQs at 40 C.F.R. Part 355, Appendix A.[6]

CERCLA Hazardous Substances

CERCLA hazardous substances include:

1. *Listed hazardous substances.* EPA publishes an alphabetical list of hazardous substances and RQs (in pounds and kilograms) in 40 C.F.R. Table 302.[7]
2. *Listed hazardous wastes.* EPA lists hazardous wastes classified pursuant to the Resource Conservation and Recovery Act (RCRA) and RQs (in pounds and kilograms) in Table 302.4. These wastes are listed numerically according to their RCRA waste-classification numbers.[8]
3. *Characteristic hazardous wastes.* Solid wastes that are not listed in Table 302.4 are hazardous wastes if they exhibit ignitable, corrosive, or reactive characteristics.[9]
4. *Toxicity Characteristic Leaching Procedure (TCLP) wastes.* Solid waste mixtures that contain hazardous constituents listed in 40 C.F.R. Section 261.24 in excess of regulatory levels are defined as "TCLP-toxic."[10]
5. *Listed radionuclides.* EPA publishes an alphabetical list of radionuclides and their RQs (in curies) in Table 302.4.[11]

▼ Determining When a Release Occurs

The term "release" includes:

any spilling, leaking, pumping, pouring, emitting, emptying, discharging, injecting, escaping, leaching, dumping, or disposing into the environment (including the abandonment or discarding of barrels, containers, and other closed receptacles) of any hazardous chemical, EHS or CERCLA hazardous substance.[12]

---

5. See Section 302(a)(2), 42 U.S.C. §11002(a)(2); 52 Fed. Reg. 13387 (April 22, 1987).
6. A list of EHSs and RQs is provided at **Appendix C**. EPA publishes and reviews the list pursuant to Section 302(a), 42 U.S.C. §11002(a).
7. A list of CERCLA hazardous substances and RQs is provided at **Appendix C**.
8. See **Appendix C**.
9. See 40 C.F.R. §§261.20-.23.
10. A copy of EPA's list of TCLP-toxic contaminants is provided at **Appendix E**.
11. See **Appendix C**. All of the five classes of substances and wastes listed above are "CERCLA hazardous substances" as that term is defined at Section 101(14), 42 U.S.C. §9601(14).
12. 40 C.F.R. §355.20.

It is important for members of the regulated community to understand that EPA's definition of release covers a range of activities that is much broader than a common-sense concept of a release. EPA classifies any discharge that is not wholly contained within buildings or structures as a release.[13] According to EPA, owners or operators must report any time an RQ or more of a hazardous substance is placed in an unenclosed structure where it becomes exposed to the environment, *regardless* of whether an RQ or more of the substance actually volatilizes into the air or migrates into surrounding water or soil.[14]

EPA expressly provides that the storage of an RQ or more of a hazardous substance in a structure or tank with a vent does not constitute a release, providing the vent is designed to prevent overpressurization or explosions.[15] Owners and operators should be aware, however, that a wide range of relatively innocuous behavior is *not* covered by exceptions, and could give rise to reporting obligations. For example, opening a container of an RQ or more of a hazardous substance outdoors (or spilling an RQ or more of a hazardous substance inside a building with an open window) could technically constitute a reportable release, even though no gas escapes into the air and no liquid migrates into the soil.

EPA Guidance

Stockpiling Material

The owner or operator of a facility that stockpiles an EHS or a CERCLA hazardous substance outdoors or in a structure that is not completely enclosed must report if the stockpiled material equals or exceeds an RQ.

EXAMPLE: A hospital stockpiles coal outdoors for use as fuel for an electrical generator. The coal contains an RQ of benzene, a CERCLA hazardous substance. The owner or operator of the hospital must report the stockpiling to the appropriate SERC and LEPC, and to the National Response Center.

continued

13. See 50 Fed. Reg. 13456 (April 4, 1985).
14. 54 Fed. Reg. 22524, 22526 (May 24, 1989).
15. Id. at 22526, Note 3.

continued

For reporting purposes, the act of stockpiling counts as a "release," regardless of how much hazardous substance actually leaches or volatilizes into the environment. The fact that the material is intentionally stored outdoors for later use is not relevant for reporting purposes, although it may be relevant to EPA's decision to respond to the release.

EPCRA HOTLINE February 1990

▼ Determining Reportable Quantities

RQs for EHSs and CERCLA Hazardous Substances

Determining RQs for releases of EHSs, listed CERCLA hazardous substances, and characteristic wastes is simple. The RQs for EHSs appear in 40 C.F.R. Part 355, Appendix A.[16] The RQs for CERCLA hazardous substances, including listed RCRA hazardous wastes and radionuclides, appear in 40 C.F.R. Table 302.4.[17] EPA has set the RQ for all characteristic wastes at 100 pounds.[18]

EPA Guidance

RQs for Characteristic Hazardous Wastes

Hazardous wastes that exhibit the characteristics of ignitability, corrosivity, or reactivity may have an RQ that is different from 100 pounds, depending on the owner/operator's knowledge of the waste.

EXAMPLE: A hazardous waste mixture is released into the environment. The waste is hazardous due to its corrosive characteristic. The owner/operator responsible for the release knows that 50 percent of the waste mixture is hydrochloric acid and

continued

16. See **Appendix C**.
17. See **Appendix C**.
18. 40 C.F.R. §302.5(b).

continued

that the waste contains no other CERCLA hazardous substances. Nor does it exhibit characteristics of ignitability or reactivity. Because the owner/operator knows the concentration of hydrochloric acid in the mixture, the release does not have to be reported unless an RQ of hydrochloric acid is released. The RQ for hydrochloric acid is 5,000 pounds. Hence the RQ for the waste mixture is 10,000 pounds.

EPCRA HOTLINE January 1991

▼ Special Reporting Rules for Mixtures and Solutions of Hazardous Substances and TCLP-Toxic Wastes

Mixtures and Solutions of EHSs

EPA has not adopted special reporting procedures for releases of mixtures and solutions that contain EHSs. A release of a mixture or solution that contains an EHS must be reported only if the released mixture contains an RQ or more of an EHS.[19]

EXAMPLE: A facility releases a ten-pound mixture containing 9.5 pounds of nonhazardous gas and .5 pounds of the nerve gas Sarin, an EHS with an RQ of one pound. The release need not be reported because the quantity of Sarin in the released mixture does not exceed its RQ.

Mixtures and Solutions of CERCLA Hazardous Substances

The RQ for mixtures and solutions that contain CERCLA hazardous substances may differ depending on whether the quantity of hazardous constituents (hazardous substances in mixtures or solutions) is known or unknown at the time of release. If the quantity of hazardous constituents in a released mixture is known, the release must be reported only if the quantity of a hazardous constituent released equals or exceeds its RQ.[20]

---

19. 40 C.F.R. §355.40(a). Under this rule, a facility owner or operator cannot be liable for failing to report the release of a mixture that contains less than an RQ of an EHS. This rule contrasts with reporting rules for releases of mixtures containing CERCLA hazardous substances. A facility owner or operator can be liable for failing to report a released mixture containing less than an RQ of a CERCLA hazardous substance if the quantity of the mixture exceeded the RQ for a hazardous constituent *and* the owner or operator did not know the quantity of hazardous substance in the mixture at the time of the release. See discussion below at p. 14 .
20. 40 C.F.R. §302.6(b)(1)(i).

EXAMPLE: A facility accidentally releases 10,000 pounds of a mixture. Chemical engineers at the facility know that the mixture contains 4,000 pounds of hydrochloric acid and 6,000 pounds of nonhazardous material. The RQ for hydrochloric acid is 5,000 pounds. The release need not be reported because the amount of hydrochloric acid released was known and did not exceed the RQ.

If the quantity of the hazardous constituent is unknown, the release must be reported if the quantity of the entire mixture equals or exceeds the RQ for the hazardous constituent.[21]

EXAMPLE: A facility accidentally releases a 10,000-pound mixture containing unknown amounts of hydrochloric acid. The release must be reported, because the quantity of the released mixture exceeds the RQ for hydrochloric acid.

If a released mixture contains unknown quantities of more than one hazardous constituent, the release must be reported if the quantity of the mixture equals or exceeds the *lowest* RQ among the hazardous constituents.[22]

EXAMPLE: A facility accidentally releases a 4,000-pound mixture containing unknown amounts of hydrochloric acid (5,000 pounds RQ) and nitric acid (1,000 pounds RQ). The release must be reported because the quantity of the released mixture exceeds the RQ for nitric acid.

TCLP-Toxic Waste Mixtures

If a solid waste mixture contains a hazardous constituent listed in 40 C.F.R. Section 261.24 Table 1 at a concentration above its regulatory level, the waste mixture is TCLP-toxic.[23] A release of a TCLP-toxic waste must be reported if the quantity of waste released equals or exceeds the RQ for a hazardous constituent.[24]

EXAMPLE: A facility accidentally releases one hundred pounds of a mixture that is TCLP-toxic for benzene. The RQ for benzene is ten pounds. The release must be reported because the quantity of the released mixture exceeds the RQ for benzene.

---

21. 40 C.F.R. §302.6(b)(1)(ii).
22. Id.
23. See **Appendix E** of this manual.
24. 40 C.F.R. §302.5(b). The use of the term "EP Toxicity" in this section refers to substances that exhibit the characteristic of toxicity according to the "extraction procedure" test. TCLP-toxic wastes also exhibit the characteristic of toxicity, but are tested using a different procedure. The method of calculating RQs for EP and TCLP-toxic wastes is identical.

If a released mixture is TCLP-toxic for more than one hazardous constituent, the release must be reported if the quantity of the mixture equals or exceeds the lowest RQ among the hazardous constituents.[25]

EXAMPLE: A facility accidentally releases five pounds of a mixture that is TCLP-toxic for benzene (10 pounds RQ) and lead (1 pound RQ). The release must be reported because the quantity of the released mixture exceeds the RQ for lead.

Mixtures of Radionuclides

The duty to report a release of a radioactive mixture may depend on:

1. Whether the identity of each radionuclide in the mixture is known; and
2. Whether the quantity of each radionuclide (in curies) in the mixture is known.

If the identity and quantity of every radionuclide in a mixture is known, the release must be reported if the ratios of quantity released (in curies) to RQ (in curies) for each radionuclide add up to one or greater.[26]

EXAMPLE: A facility accidentally releases a radioactive mixture that contains 50 curies of cesium-129 and 500 curies of cobalt-61. The RQ for cesium-129 is 100 curies and the RQ for cobalt-61 is 1,000 curies. The release must be reported because the ratio of curies of cesium-129 to its RQ (50/100 or 1/2) plus the ratio of curies of cobalt-61 to its RQ (500/1,000 or 1/2) equals 1.

If the radionuclides in a released mixture are known, but their quantities are unknown, the release must be reported if the quantity of the mixture equals or exceeds the lowest RQ among the radionuclide constituents.[27]

EXAMPLE: A facility accidentally releases 550 curies of a radioactive mixture containing unknown quantities of cesium-129 (100 curies RQ) and cobalt-61 (1,000 curies RQ). The release must be reported because the quantity of the mixture (550 curies) exceeds the RQ for cesium-129.

---

25. Id.
26. 40 C.F.R. §302.6(b)(2)(i).
27. 40 C.F.R. §302.6(b)(2)(ii).

If a released radioactive mixture contains known and unknown radionuclides, the release must be reported if the quantity of the release equals or exceeds one curie or the lowest RQ among known radioactive constituents, whichever is lower.[28]

EXAMPLE 1: A facility accidentally releases 500 curies of a radioactive mixture containing cobalt-61 (1,000 curies RQ) and other unknown radionuclides. The release must be reported because the quantity of the mixture exceeds 1 curie.

EXAMPLE 2: A facility accidentally releases .5 curie of a mixture containing plutonium-236 (.1 curie RQ) and other unknown radionuclides. The release must be reported because the quantity of the mixture exceeds the RQ for plutonium-236.

▼ Dual CERCLA and EPCRA Reporting Requirements

EPA's lists of EHSs and CERCLA hazardous substances overlap. Approximately 100 EHSs are also CERCLA hazardous substances.[29] Approximately 250 are not.[30] If a facility releases an EHS that does not require reporting under CERCLA, the facility may satisfy its reporting obligations by notifying the SERCs and LEPCs for every emergency planning district likely to be affected by the release.[31] Facilities that release substances, mixtures, or solutions that require reporting under CERCLA must notify appropriate state and local authorities *and* the National Response Center.[32] All notification must be made immediately after the discovery of a release by the fastest means available (e.g., telephone, radio, or fax).[33] Transportation-related releases can be reported by dialing "911" or a generic local emergency number.[34]

▼ Content of Notification

Notification must include the following information (to the extent known) at the time of a release:

---

28. 40 C.F.R. §302.6(b)(2)(iii).
29. See 54 Fed. Reg. 3388, 3389 (Jan. 23, 1989).
30. See 52 Fed. Reg. 13386.
31. See Section 304(a)(2), 42 U.S.C. §11004(a)(2). EPA has proposed to classify all EHSs as CERCLA hazardous substances. See 54 Fed. Reg. 3388 (Jan. 23, 1989). After this rule becomes final, every release of an EHS will require a report to the National Response Center. See also **Appendix A** for SERC contacts.
32. 40 C.F.R. §302.6(a). The telephone number for the National Response Center is 800-424-8802; in Washington, DC, 202-426-2675.
33. Section 304(b)(1), 42 U.S.C. §11004 (b)(l).
34. Id. This policy takes into account the fact that vehicle operators may not be familiar with local SERC or LEPC contacts when they are on the road. See 52 Fed. Reg. 13386.

1. The chemical name or identity of any substance involved in the release;[35]
2. An indication of whether the substance is on the list of EHSs;
3. An estimate of the quantity of any such substance that was released into the environment;
4. The time and duration of the release;
5. The medium or media into which the release occurred;
6. Any known or anticipated acute or chronic health risks associated with the emergency and, when appropriate, advice regarding medical attention necessary for exposed individuals;
7. Proper precautions to take as a result of the release, including evacuation (unless such information is readily available to the community emergency coordinator pursuant to the emergency plan); and
8. The name and telephone number of the person or persons to be contacted for further information.[36]

EPA Guidance

Content of Notification to the National Response Center

Section 103(a) of CERCLA does not specify the type of information persons must provide when notifying the National Response Center. As a matter of policy, however, EPA expects reports to the National Response Center to include the same information as reports to SERCs and LEPCs. This requirement imposes little, if any, inconvenience on members of the regulated community. It merely requires persons to call the National Response Center and report information that they have already gathered for purposes of complying with EPCRA.

EPCRA HOTLINE June 1989

---

35. EPA requires that owners and operators provide the exact chemical identity of released substances, as opposed to generic names or trade names. Emergency notification is not subject to trade secret protection. 52 Fed. Reg. 13392.
36. Section 304(b)(2)(A)-(H), 42 U.S.C. §11004(b)(2)(A)-(H); 40 C.F.R. §355(b)(2)(i)-(viii).

▼ Follow-Up Report

As soon as practicable after a release, the owner or operator of a facility must file a supplemental written report with the appropriate SERC and LEPC. The report must set forth and update all information provided in the original notice as well as any action taken to respond to and contain the release, any known or anticipated health risks associated with the release, and advice regarding the appropriate medical attention for exposed individuals.[37]

▼ Exceptions

The following releases do not have to be reported:

1. Releases that result in exposure to persons solely within the site or sites on which a facility is located;[38]
2. Federally-permitted releases;[39]
3. Applications of FIFRA-registered pesticide products;
4. Radionuclide releases that occur:
   a. naturally from land holdings, such as parks, golf courses, or other large tracts of land;
   b. naturally from the disturbance of land for purposes other than mining, such as agricultural or construction activity;
   c. from the dumping of coal and coal ash at utility and industrial facilities with coal-fired boilers; and
   d. from coal and coal ash piles at utilities and industrial facilities with coal-fired boilers.[40]

▼ Special Reporting Requirements for Releases that are Continuous and Stable in Quantity and Rate

Certain facilities that are subject to EPCRA reporting requirements may routinely release

---

37. Section 303(c), 42 U.S.C. §11004(c); 40 C.F.R. §355.40(b)(3).
38. Section 304(a)(4), 42 U.S.C. §11004(a)(4). Owners and operators should be aware that EPA has construed this exemption virtually out of existence. EPA regulations limit the exemption to releases resulting in exposure to persons solely within the boundaries of a facility, as opposed to persons on the site where a facility is located. 40 C.F.R. §355.40(2)(i). Under this regulation, exposure to any person outside of an enclosed structure would result in the loss of the exemption. As a matter of policy, EPA maintains that *potential exposure* to persons off-site is sufficient to trigger reporting obligations. According to EPA, any release into the environment above an RQ has the potential to expose persons off-site and should be reported. See 52 Fed. Reg. 13380-81.
39. Exempt releases include the discharge of air pollutants pursuant to Clean Air Act permits, the discharge of effluent pursuant to NPDES permits under the Federal Water Pollution Control Act, and the transportation, storage, and disposal of hazardous waste in compliance with RCRA waste management regulations.
40. 40 C.F.R. §355.40(a)(2).

EHSs or CERCLA hazardous substances into the environment as a part of their normal operations. For example, facilities may store hazardous substances in tanks with vents, and release material through evaporation. Other facilities may routinely pump hazardous substances through complicated systems of pipes and valves, and release material due to leaky seals and imperfect pipe connections. If regular EPCRA reporting requirements applied to these facilities, they would have to report releases frequently or risk the imposition of penalties. In order to ease the administrative burden of such frequent reporting, EPA has developed an alternative set of reporting requirements that apply to these facilities.[41]

To take advantage of the alternative reporting requirements, the person in charge of a facility must have a sound basis for concluding that the routine operation of the facility will result, or is currently resulting, in a release of hazardous substances that is continuous and stable in quantity and rate. Such a conclusion may be based on:

1. The professional judgment of a facility operator or engineer familiar with the facility, who can estimate the continuity and stability of the release; or
2. Reporting a sufficient number of releases to the National Response Center and appropriate SERCs and LEPCs to establish the continuity and stability of a release.[42]

The person in charge must then call the National Response Center and appropriate SERCs and LEPCs and provide them with the name and location of the facility and the type and quantity of material being released. The person in charge must declare that he or she is providing information as part of an initial continuous release notification report.[43]

Within thirty days of making an initial continuous release notification report, the facility must file a written notification with the EPA regional office for the geographical area where the release is occurring,[44] and with appropriate SERCs and EPCs. The written notification must include the following information:

(i) The name of the facility or vessel; the location, including the latitude and longitude, the case number assigned by the National Response Center or the Environmental Protection Agency; the Dun and Bradstreet number of the facility, if available; if a

41. The alternative reporting procedures apply to releases of CERCLA hazardous substances and EHSs. See 40 C.F.R. §302.8; 40 C.F.R. §355.40(a)(2)(iii).

42. 40 C.F.R. §302.8(d).

43. Id.

44. A list of EPA regional offices—with contacts, addresses, and telephone numbers—appears at **Appendix F**.

vessel, the port of registration of the vessel; the name and telephone number of the person in charge of the facility or vessel.

(ii) The population density within a one-mile radius of the facility or vessel, described in terms of the following ranges: 0 to 50 persons, 51 to 100 persons, 101 to 500 persons, 501 to 1,000 persons, more than 1,000 persons.

(iii) The identity and location of sensitive populations and ecosystems within a one-mile radius of the facility or vessel (e.g., elementary schools, hospitals, retirement communities, or wetlands).

(iv) For each hazardous substance release claimed to qualify for reporting under CERCLA section 103(f)(2), the following information must be supplied: (a) the name/identity of the hazardous substance; the Chemical Abstracts Service Registry Number for the substance (if available); and if the substance being released is a mixture, the components of the mixture and their approximate concentrations and quantities, by weight; (b) the upper and lower bounds of the normal range of the release (in pounds or kilograms) over the previous year; (c) the source(s) of the release (e.g., valves, pump seals, storage tank vents, stacks). If the release is from a stack, the stack height (in feet or meters); (d) the frequency of the release and the fraction of the release from each release source and the specific period over which it occurs; (e) a brief statement describing the basis for stating that the release is continuous and stable in quantity and rate; (f) an estimate of the total annual amount that was released in the previous year (in pounds or kilograms); (g) the environmental medium(s) affected by the release: (1) if surface water, the name of the surface water body; (2) if a stream, the stream order or average flowrate (in cubic feet/second) and designated use; (3) if a lake, the surface area (in acres) and average depth (in feet or meters); (4) if on or under ground, the location of public water supply wells within two miles; (h) a signed statement that the hazardous substance release(s) described is(are) continuous and stable in quantity and rate under the definitions in paragraph (a) of this section and that all reported information is accurate and current to the best knowledge of the person in charge.[45]

One year after filing the written notification, the facility must file a follow-up report with the appropriate EPA regional office, SERCs, and LEPCs. In the follow-up report, the facility must

45. 40 C.F.R. §302.8(e).

update all information provided in the initial written notification report.[46] Thereafter, the facility must continue to monitor the release and report all changes in information contained in the initial written report and follow-up reports to the appropriate EPA regional office, SERC, and LEPC.[47] The facility must also immediately report any statistically significant increase[48] in a release to the National Response Center and the appropriate SERCs and LEPCs, and must continually keep those agencies informed of any changes that occur at the facility that affect a release.[49]

---

46. 40 C.F.R. §302.8(f).
47. 40 C.F.R. §302.8(g).
48. A statistically significant increase is a release that exceeds the normal range of a continuous release as specified in the initial written notification and follow-up reports. 40 C.F.R. §302.8(b).
49. 40 C.F.R. §§302.8(g)-(i).

Chapter Three

Sections 311 and 312: Material Safety Data Sheet Submissions and Inventory Reporting

Owners or operators of facilities that are required to generate or maintain material safety data sheets (MSDSs) under OSHA's Hazardous Communication Standard must submit copies of their MSDSs to appropriate SERCs, LEPCs, and fire departments for each hazardous chemical that becomes present at a facility in excess of a threshold reporting quantity (TRQ). An MSDS describes the physical characteristics of a hazardous chemical, the hazards associated with its handling and use, and emergency procedures that employees and emergency personnel should follow when responding to accidents involving the chemical.[1] In lieu of submitting separate MSDSs for each hazardous chemical, owners and operators may submit lists of hazardous chemicals to the appropriate agencies. EPCRA also requires owners or operators of covered facilities to submit annual hazardous chemical inventory forms (Tier I and Tier II forms) to appropriate state and local organizations.

Material Safety Data Sheets

General OSHA Requirements

The Occupational Health and Safety Act requires employers to accurately complete an MSDS for each hazardous chemical they manufacture or import. OSHA also requires employers to obtain an MSDS for every hazardous chemical that they use in the workplace. Employers must make MSDSs available to workers, workers' representatives, OSHA officials, and representatives from the National Institute for Occupational Safety and Health upon request.[2]

Hazardous Chemicals

OSHA defines hazardous chemicals to include each chemical that is a physical or health hazard. A chemical is a physical hazard if:

---

1. A sample MSDS is provided at **Appendix G**.

2. See 29 C.F.R. §§1910, 1915, 1917, 1918, 1926, and 1928.

1. . . . there is scientifically valid evidence that it is a combustible liquid, a compressed gas, explosive, flammable, an organic peroxide, an oxidizer, unstable (reactive) or water reactive.[3]

A chemical is a health hazard if:

2. . . . there is statistically significant evidence based on at least one study conducted in accordance with established scientific principles that acute or chronic health effects may occur in exposed employees. The term "health hazard" includes chemicals which are carcinogens, toxic or highly toxic agents, reproductive toxins, irritants, corrosives, sensitizers, hepatotoxins, nephrotoxins, neurotoxins, agents which act on the hematopoietic system, and agents which damage the lungs, skin, eyes, or mucous membranes. . . .[4]

There is no comprehensive list of hazardous chemicals. The designation can apply to thousands of different chemicals, mixtures, and products that are used in the workplace.

Exemptions

OSHA does not require employers to prepare or make available MSDSs for:

1. RCRA hazardous wastes
2. Tobacco or tobacco products
3. Wood or wood products
4. "Articles," which are defined as manufactured items that: (a) are formed to a specific shape or design during manufacture; (b) have end-use functions dependent in whole or in part on their shape or design during end use; and (c) do not release or otherwise result in exposure to a hazardous chemical under normal conditions of use
5. Food, drugs, cosmetics, or alcoholic beverages in a retail establishment that are packaged for sale to consumers
6. Any consumer product or hazardous substance, as those terms are defined in the Consumer Product Safety Act and the Federal Hazardous Substances Act, respectively, when the employer can demonstrate that such material is used in the workplace in the

3. 29 C.F.R. §1910.1200(c).

4. Id.

same manner as in normal consumer use and that such use results in an exposure that is not greater than that experienced by consumers in terms of duration and frequency

7. Any drug, as the term is defined in the Federal Food, Drug and Cosmetic Act, when it is in its final form for direct administration to the patient.[5]

EPCRA does not require MSDS submissions for:

1. Any food, food additive, color additive, drug, or cosmetic regulated by the Food and Drug Administration

EPA Guidance

Reporting Exemption for Food Additives Regulated by the Food and Drug Administration (FDA)

The exemption applies to facilities that use a hazardous chemical in a manner that is regulated by FDA, even though the hazardous chemical is not actually added into a food product.

EXAMPLE: A facility uses chlorine (a hazardous chemical) to bleach flour. Because FDA regulates the bleaching of flour (see 27 C.F.R. Sections 137-200), the owner/operator need not submit an MSDS or Tier I or Tier II forms covering chlorine that is used for such purposes.

In general, EPA considers hazardous chemicals that are used in a manner "consistent with FDA regulations" to be exempt from Section 311/312 reporting requirements.

EPCRA HOTLINE March 1990

2. Any substance present as a solid in any manufactured item to the extent exposure to

---

5. 29 C.F.R. §1910.1200(b)(6).

the substance does not occur under normal conditions of use

3. Any substance that is used for personal, family, or household purposes, or is present in the same form and concentration as a product packaged for distribution and use by the general public

EPA Guidance

Reporting Exemption for Household Products

As long as hazardous chemicals are packaged in a form intended for distribution to the general public, a facility may accumulate an unlimited number of "packages" without having to report under Sections 311 or 312.

EXAMPLE: A wholesaler stores a large number of automobile batteries in a warehouse for sale to various retailers. Hazardous chemicals are present inside the batteries in excess of an applicable TRQ. The wholesaler does not have to report because the hazardous chemicals inside the batteries are in the same form or concentration as batteries sold for use by the general public.

EPCRA HOTLINE June 1990

4. Any substance that is used in a research laboratory or a hospital or other medical facility under the direct supervision of a technically qualified individual

EPA Guidance

Reporting Exemption for Hospitals or Medical Facilities

The owner/operator of a hospital or medical facility need not report under Sections 311 and 312 for hazardous chemicals that are stored on-site, providing the hazardous chemicals

continued

continued

are to be used for medical treatment or research purposes under the supervision of technically qualified individuals.

EXAMPLE: A hospital keeps a large supply of oxygen in a bulk storage tank for use throughout the hospital as needed. The oxygen is administered to patients and used for research purposes by doctors, nurses, and research professionals. The hospital need not include the quantity of oxygen in the tank in its TRQ calculations. Assuming no other sources of oxygen are present at the hospital, the hospital need not submit an MSDS or Tier I or Tier II forms for oxygen.

If the hospital uses oxygen from the tank for purposes not directly related to medical treatment or research, Section 311/312 requirements will apply to the portion of oxygen so used.

EPCRA HOTLINE October 1990

5. Any substance that is used in routine agricultural operations or is a fertilizer held for sale by a retailer to the ultimate customer.[6]

EPA Guidance

Agricultural Use Exemption

A retailer that does anything with hazardous chemicals other than sell them to consumers for use as fertilizer may have to comply with Section 311/312 reporting requirements.

continued

6. Section 301(e), 42 U.S.C. §11021(e).

continued

EXAMPLE: A retailer stores ammonia and phosphoric acid in large storage tanks. The retailer sells both ammonia and phosphoric acid to farmers to be used as fertilizer. The retailer also blends ammonia and phosphoric acid to form a compound that is sold to farmers as a fertilizer. The retailer must report on the portion of ammonia and phosphoric acid that it blends. The ammonia and phosphoric acid are not being held for sale, but are being held for processing to form a product that is sold as a fertilizer.

EPCRA HOTLINE June 1990

▼ Facilities Covered under EPCRA

The owner or operator of a facility that is required to prepare or make available MSDSs under OSHA must provide MSDSs or lists of chemicals covered by MSDSs to appropriate authorities within three months of the date that:

1. 10,000 pounds of a hazardous chemical become present at the facility;[7]
2. a hazardous chemical that is also an EHS becomes present at the facility in excess of 500 pounds or the TPQ for the EHS, whichever is lower;[8] or
3. an LEPC specifically requests a facility to submit an MSDS.[9]

▼ MSDS Reporting Requirements

Submission of MSDSs

The owner or operator of a covered facility must submit an MSDS for each hazardous chemical that becomes present at the facility in excess of the TRQs specified above to the SERC, the LEPC, and the fire department with jurisdiction over the facility.[10]

7. 40 C.F.R. §370.20(b)(1).
8. Id. See **Appendix C** of this manual for a list of EHSs and TPQs.
9. See 40 C.F.R. §§370.21(d); 370.20(b)(3).
10. Section 311(a), 42 U.S.C. §11021(a); 40 C.F.R. §370.21(a).

EPA Guidance

Owner/Operator Obligation to Submit MSDSs

The owner/operator of a facility is only required to comply with Sections 311 and 312 if OSHA requires the owner/operator to develop or maintain an MSDS for a hazardous chemical.

EXAMPLE: A construction company is contracted by a manufacturing company to perform work at the manufacturer's site. The construction company brings hazardous chemicals onto the site to perform its construction activities. During normal conditions of use and forseeable emergencies, only employees of the construction company will be exposed to any of the hazardous chemicals. Under this scenario, OSHA requires the construction contractor to prepare or have available MSDSs for hazardous chemicals that are brought onto the manufacturer's site to do the contracted work. The employer of the construction workers operates a facility during the construction phase, and should report on hazardous chemicals if appropriate thresholds are met. The manufacturing company is not obliged to report under Sections 311 or 312 because OSHA does not require it to produce or make available an MSDS.

EPCRA HOTLINE November 1990

Submission of Chemical Lists

In lieu of submitting an MSDS for each chemical, the owner or operator of a facility may submit a list of hazardous chemicals for which an MSDS submission would otherwise be required.[11] The list must include:

1. The chemical or common name of each hazardous chemical as provided on the MSDS; and
2. The common name of each mixture or the chemical or common name of each hazardous chemical component of a mixture.

---

11. Section 311(a)(2), 42 U.S.C. §11021(a)(2); 40 C.F.R. §370.21(b).

Facility owners and operators who choose to list hazardous chemicals must list them according to the following categories:

1. Immediate (acute) health hazard, including highly toxic, toxic, irritant, sensitizer, corrosive, and other chemicals [that cause adverse effects as a result of short term exposure]
2. Delayed (chronic) health hazard, including carcinogens and other hazardous chemicals [that cause adverse effects as a result of long-term exposure]
3. Fire hazard, including flammable, combustible liquid, pyrophoric, oxidizer chemicals
4. Sudden release of pressure, including explosive or compressed gas
5. Reactive, including unstable reactive, organic peroxide, and water-reactive.[12]

After the owner or operator of a facility submits a list, the LEPC may, at its discretion, require the facility to submit an MSDS for any chemical or mixture that appears on the list.[13]

▼ Reporting Options for Mixtures

EPCRA requires owners/operators to follow different procedures for determining whether a TRQ has been exceeded for a hazardous chemical that is a component of a mixture or mixtures, depending on whether the hazardous chemical is designated as an EHS pursuant to Section 303. Owners/operators may choose to base their TRQ calculations for a non-EHS hazardous chemical on the aggregate amount of the chemical that is present at a facility as a component of mixtures and in pure form.[14] Alternatively, they may base their TRQ calculations on the quantities of particular mixtures that are present at a facility and contain the hazardous chemical.[15] Under these options, if the aggregate amount of a chemical component or the amount of a mixture containing a hazardous chemical exceeds 10,000 ponds, the duty to report is triggered. In contrast, owners/operators must aggregate the amount of an EHS that is present at a facility in any form, including mixtures, and must report if the aggregate amount exceeds 500 pounds or the TPQ for the EHS, whichever is lower.[16]

Once they have determined that a TRQ has been exceeded, owners/operators may choose to report on mixtures or hazardous components of mixtures.[17] For example, a facility may satisfy

12. See 40 C.F.R. §370.2. For detailed definitions of hazardous chemical categories, see 29 C.F.R. §1910.1200.
13. 40 C.F.R. §370.21(d).
14. 40 C.F.R. §370.28(b)(1); 55 Fed. Reg. at 30640.
15. 40 C.F.R. §370.28(b)(2); 55 Fed. Reg. at 30640.
16. 40 C.F.R. §370.28(c).
17. 40 C.F.R. §370.28(a)(1)-(2).

its reporting obligations regarding a mixture containing hydrochloric acid by submitting an MSDS for the mixture, if one is available. Alternatively, it may submit an MSDS for hydrochloric acid.[18]

Aggregating Hazardous Chemical Components To Establish a Duty To Report

As noted above, the owner or operator of a facility may choose to aggregate the quantity of a non-EHS hazardous chemical present at a facility and report if over 10,000 pounds is present. This option allows owners and operators to keep more than 10,000 pounds of mixtures containing a hazardous chemical on-site without reporting, provided that less than 10,000 pounds of a hazardous chemical are present at the facility in the mixtures or in other forms. It should be noted that hazardous chemicals that constitute less than 1 percent (or .1 percent if carcinogenic) of a mixture are not considered to be present at a facility for reporting purposes.[19]

EXAMPLE: Three mixtures are present at a facility: (1) 10,000 pounds of Mixture A, which is 5 percent Hazardous Chemical X; (2) 20,000 pounds of Mixture B, which is 10 percent Hazardous Chemical X; and (3) 30,000 pounds of Mixture C, which is .5 percent Hazardous Chemical X. If the facility owner/operator chooses to aggregate Hazardous Chemical X, reporting is not necessary because only 2,500 pounds of Hazardous Chemical X is present at the facility for reporting purposes. Note that the 150 pounds of Hazardous Chemical X that is present in Mixture C is not counted.

Using Mixtures To Establish a Duty To Report

Instead of aggregating hazardous chemicals to establish a duty to report, an owner/operator may choose to report if 10,000 pounds of a mixture containing a hazardous chemical are present at a facility.[20] Under this option, the owner or operator of a facility may keep less than 10,000 pounds of different mixtures on-site without reporting, even if 10,000 pounds of a hazardous chemical are contained in the mixtures.

EXAMPLE: Three different mixtures are present at a facility: (1) 9,000 pounds of Mixture A, which is 33 percent Hazardous Chemical X; (2) 8,000 pounds of Mixture B, which

---

18. Id. See also 55 Fed. Reg. 30632, 30640 (July 26, 1990).
19. 40 C.F.R. §370.28(b)(1).
20. 40 C.F.R. §370.28(b)(2); 55 Fed. Reg. 30640.

is 50 percent Hazardous Chemical X; and (3) 7,000 pounds of Mixture C, which is 70 percent Hazardous Chemical X. If the owner/operator chooses to report on the mixtures—as opposed to aggregating Hazardous Chemical X—reporting is not necessary, even though 11,900 pounds of Hazardous Chemical X is present at the facility.

If the owner or operator does not know the concentration of a hazardous chemical in a mixture, EPA requires reporting on the mixture.[21] This policy makes sense, because aggregating chemical components to determine whether a TRQ has been exceeded is impossible without knowledge of the concentration of hazardous components in mixtures.

Mixtures that Contain EHSs

The owner or operator of a facility must report every time the quantity of an EHS present at the facility exceeds 500 pounds or the TPQ for the EHS, whichever is lower.[22] To determine whether a reporting threshold has been exceeded, the owner or operator must aggregate the quantity of an EHS present in mixtures at the facility with the quantity of the EHS present in other forms.[23] If the total exceeds a TRQ (500 pounds or the TPQ for the EHS, whichever is lower), the owner or operator must: (1) submit an MSDS and inventory forms covering the EHS, or (2) submit an MSDS and inventory forms for each mixture that contains an EHS, even if the quantity of each mixture present at the facility is less than the reporting threshold for the EHS.[24]

▼ Hazardous Chemical Inventory Reporting Requirements

The owner or operator of a facility that is required to submit an MSDS for a hazardous chemical must also provide the appropriate SERC, LEPC, and fire department with a Tier I or Tier II inventory form covering hazardous chemicals that are present at the facility in an amount equal to or exceeding a TRQ.[25] The Tier I form requires owners or operators to specify the quantities and general location of chemicals that present physical and health hazards.[26] The Tier II form requires owners or operators to provide more specific information regarding the

---

21. 55 Fed. Reg. at 30641.
22. 40 C.F.R. §370.38(c). See **Appendix C** to determine TPQs for EHSs.
23. 40 C.F.R. §370.28(c)(1)(i)-(ii).
24. 40 C.F.R. §370.28(c)(i)-(ii).
25. 40 C.F.R. §370.25(a)-(b).
26. A sample Tier I form, along with instructions, is provided at **Appendix H**. See also 40 C.F.R. §370.40.

identity, quantity, and location of particular hazardous chemicals on-site.[27]

Forms are due on or before March 1 of each calendar year. The owner or operator must submit a Tier I form covering chemicals that were present at the facility in excess of a reporting threshold during the preceding year.[28] The owner or operator may file a Tier II form covering specific chemicals, in lieu of a Tier I form.[29] The SERC, LEPC, or fire department may request a Tier II form for any hazardous chemical present at the facility.[30] The owner or operator of a facility must submit a completed Tier II form within thirty days of such a request.[31]

▼ Public Access to Information

Any member of the general public may request and obtain MSDSs and Tier II forms that are in the possession of SERCs and LEPCs.[32] If a state or local official acting in an official capacity requests Tier II information that is not in the hands of a SERC or LEPC, the SERC or LEPC must request the information from the facility if the information pertains to hazardous chemicals stored at the facility in excess of a TRQ.[33] If a member of the public requests and submits a general statement of need for Tier II information that is not in the hands of a SERC or LEPC, the SERC or LEPC may request the information from a facility.[34]

An owner or operator may withhold the specific identity of a hazardous chemical in order to protect a trade secret.[35] Owners and operators may also prevent disclosure of the exact location of hazardous chemicals by instructing SERCs and LEPCs on a Tier II form not to disclose exact locations to the public.[36]

---

27. A sample Tier II form, along with instructions, is provided at **Appendix I**. See also 40 C.F.R. §370.41.
28. 40 C.F.R. §370.25(a).
29. 40 C.F.R. §370.25(b).
30. These agencies have the power to require Tier II forms for all hazardous chemicals on-site, regardless of whether they were ever present at a facility in excess of a TRQ or covered in a Tier I form. See 40 C.F.R. §370.25(c), 40 C.F.R. §370(b)(3).
31. 40 C.F.R. §370.25(c).
32. 40 C.F.R. §370.31.
33. 40 C.F.R. §370.30 (b)(2).
34. 40 C.F.R. §370.30 (b)(3).
35. See Chapter Five regarding procedures for withholding information as a trade secret.
36. 40 C.F.R. §370.31.

Chapter Four

Section 313: Toxic Chemical Release Reporting

Section 313 of EPCRA requires certain manufacturing facilities to file annual release reports with EPA and appropriate state agencies for each toxic chemical that they manufacture, process, or otherwise use in excess of a threshold reporting quantity. Reports are due each July 1, and cover activities during the previous calendar year. In contrast to Section 304, which requires the reporting of *emergency* releases, Section 313 requires covered facilities to report releases, including permitted releases, that occur during normal operations. Information procured under Section 313 provides the government and the general public with an extensive body of data that can be used to determine the nature and extent of pollution problems nationally and assess the need for additional environmental regulation. Toxic release inventory information is available through the National Library of Medicine's TOXNET computer system.

Covered Facilities

Manufacturing facilities that have ten or more full-time employees[1] are subject to Section 313 reporting requirements if they manufacture or process 25,000 pounds of a toxic chemical or otherwise use 10,000 pounds of a toxic chemical in a calendar year.[2] EPA publishes a list of toxic chemicals in the Code of Federal Regulations.[3]

---

1. A facility must calculate the number of full-time employees by totaling the number of hours worked during the calendar year by all employees, including contract employees, and dividing that total by 2,000. 40 C.F.R. §372.3.
2. See Section 303(b)(1)(A), 42 U.S.C. §11023(b)(1)(A); 40 C.F.R. §372.22; Section 303(f), 42 U.S.C. §11023(f); 40 C.F.R. §372.25. The 25,000- and 10,000-pound requirements are commonly referred to as threshold reporting quantities.
3. See 40 C.F.R. §372.65. A copy of EPA's list of toxic chemicals is provided at **Appendix J**.

EPA Guidance

Calculating Numbers of Full-Time Employees

The total number of hours worked at a facility must include hours worked by every employee during a calendar year, regardless of whether they performed manufacturing tasks or worked at a facility during manufacturing operations.

EXAMPLE: A manufacturing facility shut down on January 30, 1989. Between January 1, 1989 and January 30, 1989, the facility manufactured a toxic chemical in excess of 25,000 pounds, and 10,000 hours had been worked at the facility. After the manufacturing activities ceased, six employees remained to work on wiring and warehouse activities. The facility owner/operator must add the hours worked by the six employees to the 10,000 worked before January 30. If the total number of hours exceeds 20,000, the facility must report.

EPCRA HOTLINE February 1990

Determining Whether a Facility Is a Manufacturing Facility

A facility is a "manufacturing facility" under Section 313 if it has a Standard Industrial Classification code (SIC code) beginning with 20 through 39.[4] In order to properly classify facilities according to SIC codes, one must become familiar with the distinction between "facilities" and "establishments." A *facility* includes:

> all buildings, equipment, structures, and other stationary items which are located on a single site or contiguous or adjacent sites and which are owned by the same person (or by any person which controls, is controlled by, or under common control with such person).[5]

An *establishment* includes:

4. The Office of Management and Budget (OMB) developed the SIC code system in order to provide government agencies with a uniform method for collecting and presenting business data. OMB published a comprehensive list of SIC codes in the *Standard Industrial Classification Manual 1987.* The manufacturing SIC codes that are relevant to Section 313 are provided at **Appendix K**.
5. 40 C.F.R. §372.3.

[any] economic unit, generally at a single physical location, where business is conducted or where services or industrial operations are performed.[6]

One facility can include several different establishments, each with a distinct SIC code.

EXAMPLE: A corporation owns 300 acres of land. Three buildings are located on the site: (1) meat-packing plant, (2) leather tanning and finishing facility, and (3) a small building housing a radio station. The three buildings together constitute the facility. Each building, however, is a separate establishment, and has a different SIC code depending on the type of business conducted within. The SIC code for the meat-packing plant is 2011; the SIC code for the leather tanning and finishing plant is 311; and the SIC code for the radio station is 483.

A facility has an SIC Code beginning with 20 through 39 if:

1. The facility is an establishment with a primary SIC code of 20 through 39;
2. The facility is a multi-establishment complex where all establishments have a primary SIC code of 20 through 39; or
3. The facility is a multi-establishment complex in which one of the following is true:
   a. the sum of the value of products shipped and/or produced from those establishments that have a primary SIC code of 20 through 39 is greater than 50 percent of the total value of all products shipped and/or produced from all establishments at the facility; or
   b. one establishment having a primary SIC code of 20 through 39 contributes more in terms of value of products shipped and/or produced than any other establishment within the facility.[7]

▼ Determining Threshold Planning Quantities and Reporting Obligations

Calculating Threshold Reporting Quantities

EPCRA requires owners and operators of facilities to accurately determine whether more

---

6. Id.

7. 40 C.F.R. §372.22(b).

than a TRQ of a hazardous chemical is manufactured, processed, or otherwise used during a calendar year.[8] The term *manufacture* means to:

> produce, prepare, import, or compound a toxic chemical. Manufacture also applies to a toxic chemical that is produced coincidentally during the manufacture, processing, use, or disposal of another chemical or mixture of chemicals as a byproduct, and a toxic chemical that remains in that other chemical or mixture of chemicals as an impurity.[9]

For the most part, the definition of "manufacture" is clear and straightforward. It is important to note, however, that the term includes *importing* toxic chemicals. EPA considers a facility to be an importer if it causes toxic chemicals to cross the border of the United States and enter into U.S. territory. For every shipment of chemicals that crosses the border, there can be only one importer. According to EPA, the facility that specifies the type and amount of toxic chemical to be imported is the importer, regardless of whether other parties are involved in an import transaction. Import brokers, for example, do not incur Section 313 reporting obligations by helping facilities arrange to import chemicals.[10]

The term *process* means:

> the preparation of a toxic chemical, after its manufacture, for distribution in commerce: (1) In the same form or physical state as, or in a different form or physical state from which it was received by that person so preparing such substance, or (2) As part of an article containing the toxic chemical. Processing also applies to the processing of a toxic chemical contained in a mixture or trade name product.[11]

Processing is an incorporative activity. Processing occurs when a chemical is incorporated into a product that is distributed in commerce. Incorporation can involve reactions that convert a chemical, actions that change the form or physical state of a chemical, the blending or mixing of a chemical with other chemicals, the inclusion of a chemical in an article, or the repackaging of a chemical. Examples include using chemicals as raw materials or intermediaries in the manufacture of other chemicals, the formulation of mixtures in which the incorporation of the chemical imparts some desired property to the product, and preparing chemicals for distribution in a particular form or package.[12]

---

8. 40 C.F.R. §372.30(a).
9. 40 C.F.R. §372.3.
10. 53 Fed. Reg. 4500, 4505 (Feb. 16, 1988).
11. Id.
12. 53 Fed. Reg. 4506.

EPA Guidance

Processing Toxic Chemicals

EPA does not consider the "processing" of toxic chemicals to include repackaging containers of hazardous chemicals.

EXAMPLE: A facility receives a shipment of five-gallon cans of paint containing a listed toxic chemical. The facility breaks up the shipment into separate five-gallon cans, and packages each can with a paintbrush for sale. "Processing" includes repackaging toxic chemicals. However, "repackaging" refers to the act of removing a toxic chemical from one container and placing it in another. Repackaging a *container* that contains a toxic chemical does not constitute processing.

In general, a hazardous chemical is not processed unless it is incorporated into a product for distribution in commerce.

EXAMPLE: A melamine formaldehyde resin containing a small amount of unreacted formaldehyde resin is purchased by a facility, dissolved in water, and applied to paper to produce a polymer-coated product. In the process of coating the paper, all of the formaldehyde evaporates. Because the formaldehyde is not incorporated into the final product, it is otherwise used—not processed—and the 10,000-pound TRQ applies.

EPCRA HOTLINE October 1992; March 1991

The term *otherwise use* means:

use of a toxic chemical that is not covered by the terms "manufacture" or "process" and includes use of a toxic chemical contained in a mixture or trade name product. Relabeling or redistributing a container of a toxic chemical where no repackaging of the toxic chemical occurs does not constitute use or processing of the toxic chemical.[13]

---

13. Id.

In contrast to "process," "otherwise use" is a nonincorporative activity. Facilities otherwise use toxic chemicals in activities that support, promote, or contribute to the facility's activities when the chemical does not intentionally become part of a product distributed in commerce. Examples include the use of chemicals as catalysts, solvents, reaction terminators, lubricants, refrigerants, cleaners, degreasers, and fuels.

EPA Guidance

Otherwise Using Toxic Chemicals

The term "otherwise use" encompasses a broad range of activity, including activity that is not directly related to manufacturing or processing operations.

EXAMPLE: A facility receives material X packaged in fifty-five-gallon drums. Material X is immersed in methanol, which acts as a packaging/cooling medium for material X during transport. As soon as the facility receives the delivery, employees recap the drums and send them back to the supplier. The owner/operator must report on the methanol in the drums that are opened and sent back. The methanol, in this instance, is otherwise used at the facility.

EPCRA HOTLINE September 1991

Facilities often manufacture, process, and otherwise use toxic chemicals during normal operations.

EXAMPLE: Facility A imports 25,000 pounds of Toxic Chemical X. It mixes the 25,000 pounds of Toxic Chemical X with 10,000 pounds of Toxic Chemical Y. The chemicals react to form 35,000 pounds of Toxic Chemical Z. To speed up the reaction, 5,000 pounds of Toxic Chemical C is added to the mixture to serve as a catalyst. The facility has manufactured and processed 25,000 pounds of Toxic Chemical X. It has processed 10,000 pounds of Toxic Chemical Y, manufactured 35,000 pounds of Toxic Chemical Z, and otherwise used 5,000 pounds of Toxic Chemical C.

Determining TRQs for Activities Involving Chemicals from the Same Chemical Category

EPA's list of toxic chemicals includes a list of approximately twenty toxic chemical categories.[14] The chemical categories include all compounds that contain a listed "parent" toxic chemical as part of their chemical structure. The owner or operator of a facility that manufactures, processes or otherwise uses more than a TRQ of compounds that belong to the same category must report. The weight of the compounds that contain the same parent chemical must be used to determine whether a TRQ has been exceeded. After an owner or operator determines that a TRQ has been exceeded, it need only report the weight of the parent chemical that is released during the reporting year.[15]

EXAMPLE: Compounds X and Y both contain lead as part of their chemical structures. The molecular weight of both compounds is 50 percent lead. A facility manufactured 15,000 pounds of each chemical during a calendar year. Because the compounds belong to the same chemical category and the quantity manufactured at the facility exceeded 25,000 pounds, the owner or operator must report for lead compounds. When filling out the reporting form, the owner or operator need account only for releases and transfers of lead, and may exclude the mass of other elements in Compounds X and Y from reporting calculations.

Determining TRQs for Toxic Chemicals Processed or Used in Recycling Operations

Owners and operators that process or otherwise use toxic chemicals in recycling or reuse operations must include in their TRQ calculations only the quantity of toxic chemicals they add to their operations.[16]

EXAMPLE 1: In Year 1, a facility implements a recycling operation that uses Toxic Chemical X. In that year it adds 15,000 pounds of Toxic Chemical X to the operation. The addition of 15,000 pounds of Toxic Chemical X exceeds the TRQ for use, and the owner or operator of Facility A must report.

EXAMPLE 2: In Year 2, the facility continues to run the recycling operation. It adds only 1,000 pounds of Toxic Chemical X to the system that year. Even if the total amount of

---

14. 40 C.F.R. §372.65(c). A list of toxic chemical categories is provided at **Appendix L**.

15. 40 C.F.R. §372.25(h).

16. 40 C.F.R. §372.25(e); 53 Fed. Reg. 4508.

Toxic Chemical X in the operation exceeds 10,000 pounds, the owner or operator of the facility need not report because it added less than 10,000 pounds in the calendar year.

Obligation to Report on All Activities that Involve a Toxic Chemical, Not Just Activities that Trigger a Reporting Obligation

Different activities (e.g., manufacturing, processing, or use) involving the same toxic chemical may occur at a facility during the same calendar year. The owner or operator of a facility must file a report for a toxic chemical if it manufactures over 25,000 pounds, processes over 25,000 pounds, or uses over 10,000 pounds of a toxic chemical. If a facility exceeds only one TRQ for a hazardous chemical, it must file a report that covers all activities at the facility that involve the hazardous chemical—not just the activity that exceeded the TRQ.[17]

EXAMPLE: A facility manufactured 20,000 pounds, processed 20,000 pounds, and used 11,000 pounds of Toxic Chemical X last year. Because it used more than 10,000 pounds, the facility must file a report for Toxic Chemical X. That report must include estimates of releases that occurred as a result of manufacturing and processing activities involving Toxic Chemical X, as well as releases that resulted from its use.

Reporting Rules for Specially Designated Toxic Chemicals

Special Activities Involving Toxic Chemicals

EPA may require only persons who manufacture a toxic chemical, or manufacture a toxic chemical in a particular way, to report.[18] For example, EPA requires only persons who manufacture isopropyl alcohol using the "strong acid" process to report for that chemical.[19] Facilities that manufacture more than a TRQ of a specially designated toxic chemical need report only on manufacturing activities involving the chemical.[20]

EXAMPLE: A facility manufactures 30,000 pounds of isopropyl alcohol using the strong acid process, processes 26,000 pounds, and uses 4,000 pounds. When the facility

---

17. 40 C.F.R. §372.25(c).
18. 40 C.F.R. §372.25(f).
19. See 40 C.F.R. §372.65 (**Appendix J**). The special designation appears as parenthetical information listed with isopropyl alcohol.
20. 40 C.F.R. §372.25(f).

completes a reporting form for isopropyl alcohol, it must estimate and report only those releases that occur during manufacturing operations.

Toxic Chemicals that Take Particular Forms

EPA requires reporting of certain toxic chemicals only when they exist in a particular form (e.g., fume, dust, solution, friable, and so on).[21] For example, EPA requires only persons who manufacture, process, or otherwise use zinc as a fume or a dust to report.[22] The owner or operator of a facility that manufactures, processes, or otherwise uses zinc must include only zinc fumes and dust in TRQ calculations. If the quantity of fumes and dust exceeds a TRQ, the owner or operator need report only on activities at the facility involving fumes and dust. Activities involving the toxic chemical in other forms need not be reported.[23]

▼ Special Rules for Owners and Operators of Facilities that Process or Otherwise Use Mixtures and Trade-Name Products that Contain Toxic Chemicals

TRQ Calculations and Reporting Obligations

If the owner or operator of a facility knows that a mixture[24] or trade-name product[25] contains a toxic chemical, the owner or operator must add the quantity of the toxic chemical present in the mixture or trade-name product to the total quantity of the toxic chemical manufactured, processed, or otherwise used at the facility in order to determine whether the facility has exceeded a TRQ.

EXAMPLE: A facility used 6,000 pounds of pure Toxic Chemical X as a degreaser last year. It also used 25,000 pounds of a trade-name product that was 20 percent Toxic Chemical X by weight. By using 6,000 pounds of Toxic Chemical X in pure form

---

21. 40 C.F.R. §372.25(g).
22. See 40 C.F.R. §372.65 (**Appendix J**). The special designation appears as parenthetical information listed with toxic chemicals.
23. 40 C.F.R. §372.25(g).
24. The term "mixture" means any combination of two or more chemicals, if the combination is not, in whole or in part, the result of a chemical reaction. However, if the combination was produced by a chemical reaction but could have been produced without a chemical reaction, it is also treated as a mixture. A mixture also includes any combination that consists of a chemical and associated impurities. 40 C.F.R. §372.3.
25. The term "trade-name product" means a chemical or mixture of chemicals that is distributed to other persons and that incorporates a toxic chemical component that is not identified by the applicable Chemical Abstracts Service registry number listed in 40 C.F.R. §372.66; 40 C.F.R. §372.3.

and 5,000 pounds as part of a trade-name product, the facility has exceeded the 10,000-pound TRQ for using a hazardous chemical.[26]

If a TRQ is exceeded for a toxic chemical, the owner or operator must report on all activities at the facility involving the toxic chemical in its pure form and as part of a mixture or trade-name product.

EXAMPLE: In the above example, when the facility completes and files its report on Toxic Chemical X, it must report on all activities involving Toxic Chemical X in its pure form *and* as part of the trade-name product.

Knowledge Requirement

EPCRA does not require owners and operators of facilities to test shipments of supplies they receive to determine whether they contain toxic chemicals.[27] Whether owners and operators "know" mixtures or trade-name products contain toxic chemicals depends primarily on information that they develop during the course of normal business operations, including information that their suppliers make available to them. For example, EPCRA presumes that an owner or operator knows a mixture or trade-name product contains a toxic chemical if:

1. He or she knows or has been told the Chemical Abstract Service registry number (CASRN) of the chemical and that identity or number corresponds to an identity or number on EPA's list of toxic chemicals; or
2. He or she has been told by the supplier of the mixture or trade-name product that the mixture or trade-name product contains a toxic chemical subject to Section 313 of EPCRA.[28]

Owners and operators have various reporting options depending on their knowledge of the types and concentrations of toxic chemicals contained in mixtures and trade-name products.

For example, if the owner or operator knows the specific chemical identity of the toxic chemical and the concentration at which it is present in a mixture or trade-name product, the

---

26. 40 C.F.R. §372.30(b)(1).
27. See 53 Fed. Reg. 4509-4510.
28. 40 C.F.R. §372.30(b)(2).

owner or operator must determine the quantity of the toxic chemical imported, processed, or otherwise used as part of the mixture and factor that quantity into the TRQ calculations for the facility. If the facility exceeds a TRQ, it must report on all activities involving the hazardous chemical in pure form and as part of a mixture or trade-name product.[29]

If the owner or operator of a facility knows the specific identity of a toxic chemical and has been told the upper-bound concentration at which it is present in a mixture or trade-name product, the owner or operator shall assume that the upper-bound concentration is the actual concentration when calculating whether TRQ has been exceeded. If the facility exceeds a TRQ, all activities involving the chemical in pure form and as part of a mixture or trade-name product must be reported.[30]

If the owner or operator knows the specific chemical identity of a toxic chemical in a mixture or trade-name product, but does not know its specific concentration or its upper-bound concentration and has not otherwise developed information on the composition of the chemical in the mixture or trade-name product, then it is not necessary to factor the chemical contained in the mixture or trade-name product into TRQ and release calculations.[31]

If an owner or operator knows that a mixture or trade-name product contains a toxic chemical, does not know the specific identity of the chemical, but does know its actual concentration or has been told its upper-bound concentration, the owner or operator must determine the quantity of the chemical imported, processed, or used at the facility.[32] If that quantity exceeds a TRQ, the owner or operator must report on all activities involving the hazardous chemical as part of the mixture or trade-name product.[33]

If the owner does not know the specific identity, the concentration, or the upper-bound concentration of a toxic chemical in a mixture or trade-name product, the owner or operator has no reporting obligation.[34]

▼ Toxic Chemical Release Inventory Forms

On or before July 1 of each year, the owner or operator of a facility must complete and file a Toxic Chemical Release Inventory Reporting Form (EPA Form R)[35] for each hazardous chemical that it manufactured, processed, or otherwise used in excess of a TRQ during the

29. 40 C.F.R. §372.30(b)(3)(i).
30. 40 C.F.R. §372.30(b)(3)(ii).
31. 40 C.F.R. §372.30(b)(iii).
32. When determining quantity, the owner or operator must assume that the upper-bound concentration is the actual concentration. 40 C.F.R. §372.30(b)(v).
33. 40 C.F.R. §372.30(b)(iv)-(v).
34. 40 C.F.R. §372.30(b)(vi).
35. A copy of EPA Form R is provided at **Appendix M**.

preceding calendar year.[36] The form must be submitted to EPA and appropriate state agencies.[37] In general, Form R requires the owner or operator to provide information about the facility, including its location, the type of activities carried on there, the quantity of the toxic chemical on-site, and "reasonable estimates" of the quantity of the toxic chemical released into various environmental media and transferred to treatment, storage, and disposal (TSD) facilities.

In 1991, EPA modified Form R to implement the Pollution Prevention Act of 1990. In addition to the information described above, the modified form requires facility owners and operators to provide data regarding recycling and waste reduction activities that involve the toxic chemical that is the subject of the release report. In general, the modified form requires owners or operators to provide information regarding:

1. The quantity of the chemical entering any waste stream (or otherwise released into the environment) before recycling, treatment, or disposal during the calendar year for which the report is filed and the percentage change from the previous year
2. The amount of the chemical from the facility that is recycled (at the facility or elsewhere) during each calendar year, the percentage change from the previous year, and the process of recycling used
3. The source reduction practices used with respect to the chemical during such year at the facility
4. The amount expected to be reported under 1 and 2 for the two calendar years immediately following the calendar year for which the report is filed
5. A ratio of production in the production (reporting) year to production in the previous year, or, when appropriate, an activity index based on some variable other than production, if that variable is the primary influence on waste characteristics or volume
6. The techniques that were used to identify source reduction opportunities, including but not limited to employee recommendations, external and internal audits, participative team management, and material balance audits
7. The amount of any toxic chemical released into the environment that resulted from a catastrophic event, remedial action, or other one-time event, and is not associated with production processes during the reporting year

---

36. 40 C.F.R. §372.30(d).

37. Completed forms must be sent to EPCRA Reporting Center, P.O. Box 23779, Washington, DC 20026-3779. A list of state-designated contacts is provided at **Appendix N**.

8. The amount of the chemical from the facility that is treated (at the facility or elsewhere) during the calendar year, and the percentage change from the previous year.[38]

Owners and operators of covered facilities can obtain forms and detailed reporting instructions by contacting the Section 313 Document Distribution Center, P.O. Box 12505, Cincinnati, Ohio 45212. Also, EPA has published an interim rule that provides detailed instructions for complying with the Pollution Prevention Act of 1990.[39]

Exemptions

Threshold Concentration Exemption

Facilities that manufacture, process, or otherwise use toxic chemicals that constitute less than 1 percent of a mixture (or .1 percent if carcinogenic) need not include such toxic chemicals in their TRQ calculations or release reports.[40]

EXAMPLE 1: A facility uses 100,000 pounds of a mixture containing 1 percent Toxic Chemical X and 9,500 pounds of pure Toxic Chemical X in a calendar year. The facility need not complete an EPA Form R for Toxic Chemical X. The 1,000 pounds of Toxic Chemical X present in the mixture is not included in the facility's TRQ calculation, and the 9,500 pounds of Toxic Chemical X does not exceed the appropriate TRQ (10,000 pounds).

EXAMPLE 2: A facility uses 100,000 pounds of a mixture containing 1 percent Toxic Chemical X in a calendar year. The same year, the facility uses 11,000 pounds of pure Toxic Chemical X. The facility must complete and file an EPA Form R for Toxic Chemical X because the amount used in pure form exceeded the appropriate TQ (10,000 pounds). However, when completing the form, Facility A must only account for releases of Chemical X that occur in its pure form. Facility A need not estimate and report releases of Toxic Chemical X contained in the mixture.

---

38. Section 6607 of the Pollution Prevention Act of 1990.
39. See 56 Fed. Reg. 48474 (September 25, 1991). Note that hazardous chemicals that are treated or recycled are not necessarily "released."
40. 40 C.F.R. §372.38.

EPA Guidance

De Minimis Concentrations

The owner/operator of a facility cannot avoid reporting under Section 313 by diluting a mixture to the point where toxic chemicals are present below the de minimis concentration cutoff.

EXAMPLE: A facility brings on-site a five-percent solution of Toxic Chemical X for cleaning. The cleaning solution is diluted with water and the concentration of Toxic Chemical X drops to .5 percent. The de minimis concentration exemption does not apply initially because the concentration of Toxic Chemical X was 5 percent when the solution was brought on-site. Use of the solution begins with the step of dilution. The owner/operator must count all of the Toxic Chemical X that is diluted towards the 10,000-pound TRQ for use. The owner/operator must also report releases that occur during the handling of the 5 percent solution. Once the solution is diluted, however, the owner/operator need not report releases of Toxic Chemical X that occur through the use of the .5 percent solution.

EPCRA HOTLINE June 1990

Use Exemptions

Facility owners or operators need not include toxic chemicals in TRQ calculations or release reports if they are:

1. Present in an article[41] on facility property, providing that the release of a toxic chemical does not result from the processing or use of the article;[42]

---

41. The term "article" includes "a manufactured item: 1. which is formed to a specific shape or design during manufacture; 2. which has end use functions dependent in whole or in part upon its shape or design during end use; and 3. which does not release a toxic chemical under normal conditions of processing or use of that item at the facility. 40 C.F.R. §372.3.

42. 40 C.F.R. §372.38(b).

EPA Guidance

Article Exemption

A toxic chemical contained in an article loses its exemption if more than 1/2 pound of the toxic chemical is released as a result of the use or processing of the article.

EXAMPLE: A facility cleans copper wire using a sulfuric acid solution. The solution etches away a portion of the wire. The waste stream containing the copper is sent directly to a publicly-owned treatment works (POTW) and no other releases of copper occur on-site to any other environmental media. EPA considers the transfer of a toxic substance to a POTW to be a release into the environment. According to Directive #1 in the 1989 Section 313 Reporting Package, EPA allows releases of less than 1/2 pound to be rounded to zero for the purpose of allowing toxic chemicals contained in articles to retain their exemption. Thus, if less than one pound of copper is released into the POTW, the facility does not have to include the copper present in the wire in its TRQ calculations. If more than 1/2 pound is released, the facility must count the quantity of the copper in the wire toward the total quantity of copper processed at the facility.

EPCRA HOTLINE June 1990

2. Used as a structural component of the facility;
3. Used for routine janitorial or facility grounds maintenance (examples include use of cleaning supplies, fertilizers, and pesticides similar in type or concentration to consumer products);
4. Present in products used by employees or other persons at the facility as foods, drugs, cosmetics, or other personal items, including supplies of such products within the facility such as in a facility-operated cafeteria, store, or infirmary;

EPA Guidance

Exemption for Toxic Chemicals Used To Prepare Products for Consumption by Employees

Toxic chemicals that are used to prepare items for consumption by employees need not be included in TRQ calculations or release reports.

EXAMPLE: A facility adds chlorine to its water supply system. The employees use the chlorinated water for drinking purposes. Chlorine that is used to prepare potable water for employees is exempt from Section 313 reporting requirements.

EPCRA HOTLINE June 1990

5. Present in products used for the purpose of maintaining motor vehicles operated by the facility;
6. Present in process water and noncontact cooling water as drawn from the environment or from municipal sources, or present in air used either as compressed air or as part of combustion; or
7. Manufactured, processed, or used in a laboratory at a covered facility under the supervision of a technically qualified individual.[43]

EPA Guidance

Laboratory Exemption

Toxic chemicals that are manufactured, processed, or otherwise used in laboratories are

continued

43. 40 C.F.R. §372.38(c),(d).

continued

exempt from reporting requirements only if the laboratories serve auxiliary functions to manufacturing or processing activities at the facility.

EXAMPLE: A newspaper owner/operator has a photography lab that produces pictures that appear in the newspaper. The lab does not perform product testing or analysis for the newspaper. The primary function of the lab is to develop film to be used in the newspaper. Because the lab performs activities that are essential to the manufacturing of the newspaper, toxic chemicals that are manufactured, processed, or otherwise used in the lab are not exempt from reporting requirements.

EPCRA HOTLINE January 1990

Owner and Tenant Exemptions

Owners of leased property are not obligated to file reports if their interest in their tenants' businesses is limited to the ownership of real estate.[44] Two establishments that lease property from the same owner may treat themselves as separate facilities for reporting purposes if they do not have any common corporate or business purpose.[45]

EXAMPLE: Tenant A and Tenant B lease adjacent pieces of property from A landlord. They do not have a common corporate or business interest. Tenants A and B each used 6,000 pounds of Toxic Substance X on their leasehold property last year. Without a special reporting exemption, they would be certified as two establishments of the same facility, and would have to file an EPA Form R for Toxic Chemical X because their combined use exceeded the applicable TRQ (10,000 pounds). Under the exemption, they need not report because they are treated as separate facilities and did not exceed the applicable TRQ for a facility.

---

44. 40 C.F.R. §372.38(e).
45. A common corporate business interest includes ownership, partnership, joint ventures, ownership of a controlling interest in one person by the other, or ownership of a controlling interest in both persons by a third person. 40 C.F.R. §372.38(f).

▼ Supplier Notification Requirement

To facilitate accurate reporting, EPCRA requires facilities that manufacture or process toxic chemicals and distribute mixtures or trade-name products to provide special notification to certain customers regarding the types and quantities of toxic chemicals that are present in their mixtures and trade-name products.[46] In particular, they must provide special notice to:

1. Purchasers with SIC codes 20 through 39; and
2. Purchasers who in turn may sell or otherwise distribute mixtures or trade-name products to facilities with SIC codes 20 through 39.[47]

EPA Guidance

Supplier Notification

Supplier notification requirements apply only to owner/operators of facilities that manufacture or process toxic chemicals.

EXAMPLE: A facility uses hydrochloric acid to clean reaction vessels. The same facility also buys hydrochloric acid solution (bought as "Trade-Name X") and resells it to other customers (no repackaging or relabeling of the solution occurs). The owner/operator need not develop supplier notification for the hydrochloric acid that it sells. If the facility receives supplier notification from the manufacturer of Trade-Name X, EPA encourages the facility to pass on the notification to subsequent purchasers.

EPCRA HOTLINE August 1990

Notification shall be in writing and shall include:

1. A statement that the mixture or trade-name product contains a toxic chemical or chemicals subject to the reporting requirements of Section 313 of Title III of the

---

46. 40 C.F.R. §372.45.

47. 40 C.F.R. §372.45(a).

Superfund Amendments and Reauthorization Act of 1986 and 40 C.F.R. Part 372;

2. The name of each toxic chemical, and the associated Chemical Abstracts Service registry number if applicable as set forth in 40 C.F.R. Section 372.65; and
3. The percent by weight of each toxic chemical in the mixture or trade-name product.[48]

Suppliers must provide the above notice with the first shipment to each customer for each calendar year beginning July 1.[49] If a supplier changes the percentage of a toxic component or adds or subtracts a toxic substance from a mixture or trade-name product, the supplier must provide a revised notice reflecting the changes with the first shipment of the changed mixture.[50] The supplier must also provide purchasers with revised notices within thirty days of discovering inaccurate information in prior notices regarding the types and percentages of toxic chemicals in mixtures or trade-name products.[51] If an MSDS is required to be prepared and distributed with a mixture or trade-name product, the toxic chemical notice shall be provided with or incorporated into the MSDS.[52]

Suppliers need not provide notice if their mixtures or trade-name products:

1. Contain less than 1 percent of a toxic chemical or .1 percent of a toxic chemical that is a carcinogen;[53]
2. Are articles;[54]
3. Are foods, drugs, cosmetics, alcoholic beverages, tobacco, or tobacco products packaged for distribution to the general public; or
4. Are consumer products packaged for distribution to the general public.[55]

Suppliers may substitute a generic name for a toxic chemical if they consider the specific identity of the chemical to be a trade secret under 29 C.F.R. Section 1910.1200.[56] Suppliers may also substitute an upper-bound concentration for the actual concentration of a toxic chemical if they consider the actual concentration to be a trade secret under applicable state law or the Restatement of Torts.[57]

---

48. 40 C.F.R. §372.45(b)(1)-(3). A sample notification letter is provided at **Appendix O**.
49. 40 C.F.R. §372.45(c)(1), (2).
50. 40 C.F.R. §372.45(c)(3).
51. 40 C.F.R §372.45(c)(4)
52. 40 C.F.R. §372.45(c)(5). A sample MSDS notification is provided at **Appendix P**.
53. See **Appendix J** of this manual for a list of appropriate de minimis concentrations for toxic chemicals.
54. See note 41 for a definition of "article."
55. 40 C.F.R. §372.45(d).
56. 40 C.F.R. §372.45(e).
57. 40 C.F.R §372.45(f). See *Restatement of Torts*, Section 757, Comment b (establishing criteria for trade-secret claims).

Chapter Five

Trade Secrets

General Procedures

An owner or operator of a facility that is required to provide information under EPCRA's threshold planning, MSDS, Tier II, or Toxic Chemical Release Inventory reporting requirements may refuse to disclose the precise chemical identity of an EHS, a hazardous chemical, or a toxic chemical on the grounds that it is a trade secret.[1] When an owner or operator invokes trade secret protection, a generic chemical identity must be provided in lieu of a specific chemical identity whenever EPCRA requires reporting on a particular chemical. A generic chemical identity includes any generic chemical class or category that is descriptive of the structure of the chemical.[2] For example, an owner or operator could list "toluene diisocyanate" as "organic diisocyanate."[3]

When an owner or operator submits a generic chemical identity in a report to a SERC, LEPC, local agency, or EPA, a trade secrecy claim must also be filed with the Title III Reporting Center, Environmental Protection Agency, P.O. Box 23779, Washington, DC 20026-3779. The claim must include a written substantiation stating facts sufficient to support the following conclusions:

1. The submitter has not disclosed the information to any other person, other than a member of an LEPC, an officer or employee of the United States or a state or local government, an employee of such person, or a person who is bound by a confidentiality agreement, and such person has taken reasonable measures to protect the confidentiality of such information and continues to take such measures;
2. The information is not required to be disclosed, or otherwise made available, to the public under any other federal or state law;

---

1. 40 C.F.R. §350.5(a).
2. 40 C.F.R. §350.5(f).
3. See 40 C.F.R. §370.41, Instructions for Completing Tier II Forms.

3. Disclosure of the information is likely to cause substantial harm to the competitive position of such person; and
4. The chemical identity is not readily discoverable through reverse engineering.[4]

The submitter may provide the above information on an EPA substantiation form.[5] The submitter must also supply supplemental information to EPA upon request.[6]

After an owner or operator files a trade secrecy claim, the specific identity of a chemical is treated as confidential until a contrary determination is made.[7] When EPA receives a claim, it may:

1. Choose not to contest the validity of the claim;
2. Investigate and determine the validity of the claim *sua sponte*; or
3. Investigate and determine the validity of the claim in response to a public petition requesting disclosure of the chemical identity.[8]

EPA must determine the validity of claims within thirty days of the receipt of a public petition.[9] Parties may appeal determinations to EPA's Office of General Counsel and to a U.S. district court if necessary.[10]

▼ Special Requirements for Disclosure to Health Professionals

Nonemergencies

Owners and operators must disclose the specific identity of a chemical to a health professional on request if:

1. The request is in writing
2. The request describes why the health professional has a reasonable basis to suspect that: (a) the specific chemical identity is needed for purposes of diagnosis or treatment of an individual, (b) the individual or individuals being diagnosed or treated have been

---

4. 40 C.F.R. §350.13.
5. See 40 C.F.R. §350.27. A copy of EPA's substantiation form is provided at **Appendix Q**.
6. 40 C.F.R. §350.7(e).
7. 40 C.F.R. §350.9(a).
8. 40 C.F.R. §350.9(b). See also 40 C.F.R. §350.15 (granting the public the right to request disclosure of chemical identities).
9. 40 C.F.R. §350.11(a).
10. 40 C.F.R. §350.11(a)(2)(i); 40 C.F.R. §350.11(b).

exposed to the chemical concerned, and (c) knowledge of the specific chemical identity of such chemical will assist in diagnosis and treatment

3. The request contains a confidentiality agreement[11]

Owners and operators must also disclose specific chemical identities to local government employees conducting research on the health consequences of exposure to chemicals. Such requests must be in writing and must include confidentiality agreements.[12]

Emergencies

Owners and operators must immediately disclose specific chemical identities to any treating physician or nurse who requests such information if the treating physician or nurse determines that:

1. A medical emergency exists as to the individual or individuals being diagnosed or treated;
2. The specific chemical identity of the substance concerned is necessary for or will assist in emergency or first-aid diagnosis or treatment; and
3. The individual or individuals being diagnosed or treated have been exposed to the chemical concerned.[13]

The owner or operator must provide information immediately, but may require a written statement of need and a confidentiality agreement from the treating physician or nurse as soon as circumstances permit.[14]

---

11. Section 323(a), 42 U.S.C. §11043(a); 40 C.F.R. §350.40(c).
12. Section 323(c), 42 U.S.C. §11043(c); 40 C.F.R. §350.40(d).
13. Section 323(b), 42 U.S.C. §11043 (b); 40 C.F.R. §350.40(e) (i)-(iii) .
14. 40 C.F.R. §350.40(e) (2) .

Chapter Six

Enforcement

Like all federal health, safety, and environmental protection statutes, EPCRA contains enforcement provisions designed to deter violations. EPCRA's enforcement provisions provide EPA, SERCs, LEPCs, and private citizens with civil causes of action against violators. EPCRA establishes criminal penalties for persons who knowingly violate emergency release reporting requirements. The federal government may also criminally prosecute persons for fraudulently providing EPA with incorrect information pertaining to threshold planning, toxic chemical release reporting, and hazardous chemical inventory reporting.

▼ EPA Enforcement Authority

Civil Actions

EPCRA authorizes EPA to impose civil penalties of up to $25,000 for each day of violation on persons who violate the emergency planning requirements of Section 302, the emergency release reporting requirements of Section 304, the Tier I/ Tier II hazardous chemical inventory reporting requirements of Section 312, or the toxic chemical release reporting requirements of Section 313.[1] EPCRA authorizes EPA to impose civil penalties of up to $10,000 for each day of violation on persons who fail to comply with the MSDS reporting requirements of Section 311.[2] EPA may assess penalties only after providing alleged violators notice and an opportunity for a hearing before an administrative law judge.[3]

1. See Section 325(a), 42 U.S.C. §11045(a); Section 325(b), 42 U.S.C. §11045(b); Section 325(c)(1), (3), 42 U.S.C. §11045(c)(1),(3).
2. Section 325(c)(2), (3), 42 U.S.C. §11045(c)(2), (3).
3. The procedures for providing notice and conducting hearings appear at 40 C.F.R. Part 22.

Penalty Policy for Sections 302, 304, 311, and 312

EPA has developed a detailed internal penalty policy that is designed to adjust the severity of penalties to reflect the severity of violations.[4] EPA uses a two-step process to assess penalties. First, it assesses the acts that constitute a violation and determines a base penalty. Next, it takes into account certain characteristics of the violator to arrive at a final penalty. EPA calculates base penalties according to the following four variables:

1. *Nature of the violation:* If a person violates Sections 302, 304, or 312, the maximum penalty is $25,000. If a person violates Section 311, the maximum penalty is $10,000.
2. *Extent of the violation:* EPA determines how late a party submitted a release report, threshold planning notification, MSDS, or inventory form. The later the submission, the greater the penalty.
3. *Gravity of the violation:* EPA determines the quantity of hazardous substance released from a facility or present at a facility at the time a person failed to report or reported late. EPA increases the penalty depending on the quantity of hazardous substance released above an RQ or the quantity of hazardous substance present on-site in excess of a TPQ or TRQ.
4. *Circumstances of the violation:* EPA may adjust a penalty to reflect the extent to which a violation disrupted EPA's ability to plan for an emergency, or respond to an emergency, or resulted in health risks to the surrounding community.[5]

EPA then takes the following four variables into account in order to adjust the base penalty:

1. *Violator's ability to pay:* EPA may adjust the base penalty downward if a violator demonstrates that it lacks the ability to pay. In general, EPA will not assess penalties that are so severe that they cause a company to go out of business. However, EPA will make exceptions to this general rule in cases of repeat violators.
2. *Violator's history of previous violations:* EPA may adjust a penalty upward to reflect a person's history of past violations.
3. *Violator's culpability:* EPA may adjust a penalty upward or downward depending on a violator's culpability. Violations that occur as a result of good-faith mistakes or oversight may be penalized less severely than intentional violations.

---

4. See EPA Final Penalty Policy for Sections 302, 303, 304, 311, and 312 of EPCRA. A copy of the Penalty Policy is provided at **Appendix R**.

5. The procedure for calibrating penalties is outlined in detail at pp. 8-20 of the Penalty Policy. See **Appendix R**.

4. *The economic benefit to the violator for noncompliance:* EPA may increase a penalty to the point at which it eliminates a violator's economic incentive to violate the law.[6]

Since 1989, EPA has initiated approximately 130 enforcement actions for violations of Sections 302, 304, 311, and 312, and has collected penalties as high as $130,000 from individual violators.[7]

EPA uses the procedures outlined above to calculate single-day penalties for violations. EPA reserves multiday penalties for more serious violations and to encourage violators to return to compliance. For example, if EPA assesses a penalty of $25,000 against a company for failing to provide a follow-up release report under Section 304, the company may elect to pay the $25,000 and not report. To encourage reporting, EPA may order the company to report and impose a separate penalty for *each day* that a facility does not report after receiving the order.[8]

Special Enforcement Provisions for Section 304

To create incentives to provide the earliest possible warning in the event of an emergency release of hazardous substances, EPCRA authorizes EPA to impose especially stringent penalties on persons who violate reporting requirements under Section 304. In cases of second and subsequent violations of Section 304, EPCRA authorizes EPA to impose a base penalty of $75,000 per violation for each day that the violation continues.[9] EPCRA also authorizes criminal penalties for certain Section 304 violations, as discussed below.

Penalty Policy for Section 313

EPA has developed a penalty policy to deal exclusively with violations of Toxic Chemical Release Inventory Reporting requirements under Section 313 (the "Section 313 Policy").[10] Under the Section 313 Policy, EPA uses a two-step approach for calculating penalties. First it calculates a base penalty, then it adjusts the base penalty to arrive at a final assessment. EPA considers the following factors when calculating the base penalty:

1. *Circumstances of the violation:* EPA ranks classes of violations according to severity. Failing to submit a report more than one year after it is due is the most severe violation.

---

6. The procedures for adjusting base penalties are described in detail at pp. 22-28 of the Penalty Policy. See **Appendix R**.
7. A list of EPA enforcement actions initiated since 1989 is provided at **Appendix S**.
8. See Penalty Policy pp. 20-21.
9. Section 345(b)(2), 42 U.S.C. §11045(b)(2).
10. See *Enforcement Response Policy for Section 313 of EPCRA* (Aug. 10, 1992) (Section 313 Policy). A copy of the Section 313 Policy is provided at **Appendix T**.

Making a data quality error that is disclosed within four months of the date a report is due is among the least severe violations.

2. *The amount of toxic chemicals manufactured, processed, or otherwise used at a facility in excess of a TRQ:* EPA imposes higher base penalties on facilities that manufacture, process, or otherwise use more than ten times the reporting threshold for a toxic chemical.
3. *The size of the facility:* EPA imposes larger penalties upon larger facilities. EPA determines the size of a violator by calculating the violator's total sales and number of employees.[11]

EPA takes the following factors into account when adjusting the base penalty:

1. *Voluntary disclosure:* EPA may reduce the base penalty up to 50 percent if violations are disclosed.
2. *Prior violations:* EPA may increase the base penalty up to 100 percent for repeat violators.
3. *Delisted toxic chemicals:* EPA may settle litigation over reporting violations involving delisted chemicals for 25 percent less than the base penalty. To qualify for a reduction, a toxic chemical must be delisted before or during the pendency of an enforcement action.
4. *Violator's attitude:* EPA may adjust the base penalty upward or downward up to 30 percent depending on the violator's cooperation with EPA's investigation and good-faith efforts to comply with EPCRA.
5. *Ability to pay:* Normally, EPA will not impose on a violator a penalty that would drive the violator out of business. The violator has the burden of persuading EPA that it cannot afford to pay a penalty.
6. *Considerations of justice:* EPA will consider, on a case-by-case basis, unusual circumstances that might merit the reduction of a base penalty.[12]

As of September 1992, EPA had initiated approximately 650 enforcement actions for violations of Section 313, and had imposed penalties as high as $140,000 on individual

---

11. Detailed procedures for determining base penalties appear at pages 7 through 11 of the Section 313 Policy. See **Appendix T**.

12. Detailed procedures for adjusting base penalties appear at pp. 13-20 of the Section 313 Policy. See **Appendix T**.

violators.[13] Members of the regulated community should expect EPA's enforcement of Section 313 to become increasingly strict. A recent report by the General Accounting Office urged EPA to step up enforcement efforts and recommended that Congress grant EPA more extensive authority to enforce EPCRA's toxic release reporting requirements.[14]

Other Violations

EPA may impose civil penalties up to $25,000 for each day of violation on persons who fail to disclose trade secret information to a medical professional under Section 323(b).[15] EPA may also impose civil penalties of up to $25,000 for submitting frivolous trade secret claims under Section 324(d).[16] EPA has not developed detailed policies for calculating penalties for either of the above violations.

Criminal Enforcement

EPCRA authorizes the criminal prosecution of persons who knowingly violate the emergency release reporting requirements of Section 304. If convicted, violators face up to $25,000 in criminal fines and up to two years in prison (or in the case of a second or subsequent conviction, up to $50,000 and five years in prison).[17] EPCRA also authorizes the criminal prosecution of persons who knowingly and willfully disclose information entitled to protection as a trade secret.[18] If convicted, violators face up to $20,000 in fines and one year in prison.

EPCRA does not specifically authorize criminal prosecution for violations other than those discussed above. However, the federal government may criminally prosecute facility owner/operators under the general provisions of Section 1001 of Title 18, which penalizes persons who knowingly provide the government with fraudulent information. The government has successfully used this strategy in prosecutions under TSCA[19] and EPA has expressed a willingness to employ it when owner/operators intentionally provide inaccurate information or fail to report under EPCRA.[20] If convicted under Title 18, violators face fines of up to $10,000 and imprisonment of up to five years, or both.[21]

---

13. A list of EPA Section 313 enforcement actions can be obtained by contacting the EPCRA hotline in Washington, DC at 1-800-535-0202.
14. See GAO, *Report to Congress: EPA's Toxic Release Inventory Is Useful but Can Be Improved* (June 1991).
15. Section 325(c)(2), (3), 42 U.S.C. §11045(c)(2), (3).
16. Section 325(d)(1), 42 U.S.C. §11045(d)(1).
17. Section 325(b)(4), 42 U.S.C. §11045(b)(4).
18. Section 354(d)(2), 42 U.S.C. §11045(d)(2).
19. See, e.g., *United States v. Electric Services Co.*, No. 85-0117 (S.D. Ohio, April 4, 1986) (applying 18 U.S.C. §371, which makes it a criminal offense to conspire to provide the government with fraudulent information).
20. See Section 313 Policy at p. 7, **Appendix T**. See also, 54 Fed. Reg. 3388, 3391 (Jan. 23, 1989).
21. 18 U.S.C. §§371, 1001.

▼ State and Local Enforcement

A state or local government may commence a civil action against a facility owner/operator for failure to:

1. Notify the appropriate SERC that a facility is subject to threshold planning requirements pursuant to Section 302;
2. Comply with information requests from LEPCs pursuant to Section 303(d);
3. Submit MSDSs or Tier I or Tier II forms pursuant to Sections 311 and 312.[22]

EPCRA authorizes state and local plaintiffs to impose on violators the full range of civil penalties available to EPA.[23] The penalties are payable to the U.S. Treasury, not to the plaintiffs. EPCRA also authorizes state and local plaintiffs to recover litigation costs from violators if they are successful.[24] State and local plaintiffs may not bring suit, however, if EPA is diligently pursuing an administrative or civil action to impose a penalty or enforce compliance.[25]

▼ Citizen Suits

A private citizen may commence a civil action against the owner/operator of a facility for failure to:

1. Submit an emergency follow-up notice under Section 304;
2. Submit MSDSs and Tier I forms pursuant to Sections 311 and 312;
3. Complete and submit a Toxic Chemical Release Inventory Reporting Form (EPA Form R) pursuant to Section 313.[26]

EPCRA authorizes private plaintiffs to impose on violators the full range of civil penalties available to EPA.[27] The penalties are payable to the U.S. Treasury, not to the plaintiffs. EPCRA also authorizes private plaintiffs to recover litigation costs from violators if they are successful.[28] Private plaintiffs must provide EPA with notice of alleged violations at least sixty days prior to commencing litigation[29] and may not bring suit if EPA is diligently pursuing an administrative or civil action to impose a penalty or enforce compliance.[30]

---

22. See Section 346(a)(2), 42 U.S.C. §11046(a)(2).
23. See Section 346(c), 42 U.S.C. §11046(c).
24. Section 346(f), 42 U.S.C. §11046(f).
25. Section 346(e), 42 U.S.C. §11046(e).
26. Section 346(a)(1), 42 U.S.C. §11046(a)(1).
27. See Section 346(c), 42 U.S.C. §11046(c).
28. Section 346(f), 42 U.S.C. §11046(f).
29. Section 346(d)(1), 42 U.S.C. §11046(d)(1).
30. Section 346(e), 42 U.S.C. §11046(e).

Appendix A

State Emergency Response Commissions (SERCs)

STATE EMERGENCY RESPONSE COMMISSIONS (SERCS)

ALABAMA	Chair	302	304	311/312	313
ALABAMA EMERGENCY RESPONSE COMMISSION					
DIRECTOR, ALABAMA EMERGENCY MANAGEMENT AGENCY					
520 SOUTH COURT STREET					
MONTGOMERY, AL 36130					
PHILLIP HICKS	*				
CO-CHAIR					
Contact(s):					
DAVE WHITE					
(205) 834-1375					
ALABAMA EMERGENCY RESPONSE COMMISSION					
DIRECTOR, ALABAMA DEPARTMENT OF ENVIRONMENTAL MANAGEMENT					
1751 CONGRESSMAN W.L. DICKINSON DRIVE					
MONTGOMERY, AL 36109					
(205) 271-7700					
LEIGH PEGUES	*				
CO-CHAIR					
Contact(s):					
LEIGH PEGUES					
ALABAMA EMERGENCY RESPONSE COMMISSION					
ALABAMA DEPARTMENT OF ENVIRONMENTAL MANAGEMENT					
1751 CONGRESSMAN W.L. DICKINSON DRIVE					
MONTGOMERY, AL 36109					
(205) 260-2700					
24-Hour Emergency Number: (205) 834-1375, (205) 271-7700, (800) 843-0699					
E. JOHN WILLIFORD					
CHIEF OF OPERATIONS					
Contact(s):					
L.G. LINN		*	*	*	*
(205) 260-2700					
KEITH CARTER					
(205) 260-2700					

STATE EMERGENCY RESPONSE COMMISSIONS (SERCS)

	Chair	302	304	311/312	313
ALASKA					
ALASKA STATE EMERGENCY RESPONSE COMMISSION					
SPILL PREVENTION PLANNING & MANAGEMENT SECTION					
410 WILLOUGHBY					
SUITE 302					
JUNEAU, AK 99801-1795					
(907) 465-5220					
JOHN A. SANDER	*				
CHAIRMAN					
Contact(s):					
AMY SKILBRED		*	*	*	*
(907) 465-5220					
CAMILLE STEPHENS					
(907) 465-5239					
AMERICAN SAMOA					
PROGRAM COORDINATOR FOR THE TERRITORIAL					
EMERGENCY MANAGEMENT COORDINATION OFFICE					
AMERICAN SAMOAN GOVERNMENT					
P.O. BOX 1086					
PAGO PAGO, AS 96799					
INTERNATIONAL NUMBER (684) 633-2331					
24-Hour Emergency Number: NONE					
MAIAVA O. HUNKIN	*				
Contact(s):					
MAIAVA O. HUNKIN		*	*		
AMERICAN SAMOA EPA					
OFFICE OF THE GOVERNOR					
PAGO PAGO, AS 96799					
INTERNATIONAL NUMBER (684) 633-2304					
DIRECTOR					
Contact(s):					
PATI FAIAI				*	*

STATE EMERGENCY RESPONSE COMMISSIONS (SERCS)

	Chair	302	304	311/312	313
ARIZONA					
ARIZONA EMERGENCY RESPONSE COMMISSION					
DIVISION OF EMERGENCY SERVICES, BUILDING 341					
5636 EAST MCDOWELL ROAD					
PHOENIX, AZ 85008					
(602) 231-6326					
24-Hour Emergency Number: (602) 257-2330					
WILLIAM D. LOCKWOOD	*				
CHAIRMAN					
Contact(s):					
ETHEL DeMARR		*	*	*	*
(602) 231-6326					
ARKANSAS					
STATE EMERGENCY RESPONSE COMMISSION					
C/O DEPARTMENT OF POLLUTION CONTROL AND ECOLOGY					
P.O. BOX 8913					
LITTLE ROCK, AR 72219-8913					
(501) 562-7444					
24-Hour Emergency Number: (501) 374-1201					
RANDALL MATHIS	*				
CHAIRMAN					
Contact(s):					
JOHN WARD		*	*		
(501) 562-7444					
ATTN: SARA DIVISION					
ARKANSAS DEPARTMENT OF LABOR					
10421 WEST MARKHAM					
LITTLE ROCK, AR 72205					
(501) 562-7444					
Contact(s):					
JOHN WARD				*	*
(501) 562-7444					

09/16/92

STATE EMERGENCY RESPONSE COMMISSIONS (SERCS)

	Chair	302	304	311/312	313
CALIFORNIA					
CHEMICAL EMERGENCY PLANNING AND RESPONSE COMMISSION 2800 MEADOWVIEW ROAD SACRAMENTO, CA 95832 (916) 427-4287 RICHARD ANDREWS CHAIRMAN Contact(s): PAUL PENN	*				
CALIFORNIA EMERGENCY PLANNING AND RESPONSE COMMISSION OFFICE OF EMERGENCY SERVICES HAZARDOUS MATERIALS DIVISION 2800 MEADOWVIEW ROAD SACRAMENTO, CA 95832 (916) 427-4287 24-Hour Emergency Number: (800) 852-7550 Contact(s): TITLE III SUBMISSIONS		*	*	*	
CALIFORNIA ENVIRONMENTAL PROTECTION AGENCY 555 CAPITOL MALL SACRAMENTO, CA 95814 (916) 324-9924 ATTN: SECTION 313 REPORTS Contact(s): STEVE HANNA					*
COLORADO					
COLORADO EMERGENCY PLANNING COMMISSION COLORADO DEPARTMENT OF HEALTH 4210 EAST 11TH AVENUE DENVER, CO 80220 (303) 331-4830 DAVID C. SHELTON CHAIRMAN Contact(s): JUDY WADDILL	*				

STATE EMERGENCY RESPONSE COMMISSIONS (SERCS)

COLORADO EMERGENCY PLANNING COMMISSION
COLORADO DEPARTMENT OF HEALTH
4210 EAST 11TH AVENUE
DENVER, CO 80220
(303) 331-4858
24-Hour Emergency Number: (303) 331-4858, 377-6326 AFTER HOURS & WEEKENDS
Contact(s):

	Chair	302	304	311/312	313
PAMELA HARLEY		*		*	*

(303) 331-4858

COLORADO EMERGENCY PLANNING COMMISSION
COLORADO DEPARTMENT OF HEALTH
4210 EAST 11TH AVENUE
DENVER, CO 80220
(303) 331-4830
Contact(s):

	Chair	302	304	311/312	313
STEVE GUNDERSON			*		

(303) 331-4869

CONNECTICUT

STATE EMERGENCY RESPONSE COMMISSSION
DEPUTY COMMISSIONER, DEPARTMENT OF ENVIRONMENTAL PROTECTION
STATE OFFICE BUILDING, ROOM 115
165 CAPITOL AVENUE
HARTFORD, CT 06106
(203) 566-2110

	Chair	302	304	311/312	313
ROBERT E. MOORE	*				

CHAIRMAN
Contact(s):
ROBERT E. MOORE

RIGHT-TO-KNOW PROGRAM COORDINATOR
STATE EMERGENCY RESPONSE COMMISSION
STATE OFFICE BUILDING, ROOM 146
165 CAPITOL AVENUE
HARTFORD, CT 06106
(203) 566-4856
24-Hour Emergency Number: (203) 566-3338
EMERGENCY PLANNING AND COMMUNITY
Contact(s):

	Chair	302	304	311/312	313
SUZANNE VAUGN		*	*	*	*

STATE EMERGENCY RESPONSE COMMISSIONS (SERCS)

	Chair	302	304	311/312	313
DELAWARE					
CHAIRMAN OF SERC					
P.O. BOX 818					
DOVER, DE 19903					
(302) 739-4321					
PATRICK W. MURRAY	*				
SECRETARY DEPT. OF PUBLIC SAFETY					
Contact(s):					
DR. GORDON HENDERSON					
(302) 834-4531					
DIVISION OF EMERGENCY PLANNING AND OPERATIONS					
P.O. BOX 527					
DELAWARE CITY, DE 19706					
(302) 834-4531					
ACTING DIRECTOR					
Contact(s):					
JAMES W. HOFFMAN		*			
DIVISION OF AIR AND WASTE MANAGEMENT					
DEPARTMENT OF NATURAL RESOURCES AND ENVIRONMENTAL CONTROL					
RICHARDSON AND ROBBINS BUILDING - 89 KINGS HIGHWAY					
P.O. BOX 1401					
DOVER, DE 19903					
(302) 739-4791					
24-Hour Emergency Number: 800-622-8802					
DIRECTOR					
Contact(s):					
ROBERT PRITCHETT			*	*	*
DISTRICT OF COLUMBIA					
EMERGENCY RESPONSE COMMISSION FOR TITLE III					
OFFICE OF EMERGENCY PREPAREDNESS					
FRANK REEVES CENTER FOR MUNICIPAL AFFAIRS					
2000 14TH STREET, NW					
WASHINGTON, DC 20009					
(202) 727-6161					
PAMELA THUBER	*				
DIRECTOR					
Contact(s):					
PAMELA THUBER					*

STATE EMERGENCY RESPONSE COMMISSIONS (SERCS)

	Chair	302	304	311/312	313
EMERGENCY RESPONSE COMMISSION FOR TITLE III					
OFFICE OF EMERGENCY PREPAREDNESS					
FRANK REEVES CENTER FOR MUNICIPAL AFFAIRS					
2000 14TH STREET, NW					
WASHINGTON, DC 20009					
(202) 727-6161					
24-Hour Emergency Number: (202) 727-6161					
ENVIRONMENTAL PLANNING SPECIALIST					
Contact(s):					
PAMELA THURBER		*	*	*	

FLORIDA

	Chair	302	304	311/312	313
FLORIDA EMERGENCY RESPONSE COMMISSION					
SECRETARY, FLORIDA DEPARTMENT OF COMMUNITY AFFAIRS					
2740 CENTERVIEW DRIVE					
TALLAHASSEE, FL 32399-2100					
(904) 488-1472					
24-Hour Emergency Number: (904) 488-1320					
WILLIAM E. SADOWSKI	*				
CHAIRMAN					
Contact(s):					
EVE RAINEY		*	*	*	*
(800) 635-7179 (IN FL)					

GEORGIA

	Chair	302	304	311/312	313
GEORGIA EMERGENCY RESPONSE COMMISSION					
GEORGIA DEPARTMENT OF NATURAL RESOURCES					
205 BUTLER STREET, SE					
FLOYD TOWERS EAST, SUITE 1252					
ATLANTA, GA 30334					
(404) 656-3500					
JOE TANNER	*				
CHAIRMAN					
Contact(s):					
JOE TANNER					

STATE EMERGENCY RESPONSE COMMISSIONS (SERCS)

	Chair	302	304	311/312	313
GEORGIA EMERGENCY RESPONSE COMMISSION					
205 BUTLER STREET, SE					
FLOYD TOWER EAST, SUITE 1166					
ATLANTA, GA 30334					
(404) 656-6905					
24-Hour Emergency Number: (800) 241-4113 IN AND OUT OF GA					
Contact(s):					
BURT LANGLEY		*	*	*	*
JIM SETSER (404) 656-4713					
GUAM					
GUAM STATE EMERGENCY RESPONSE COMMISSION					
CIVIL DEFENSE - GUAM EMERGENCY SERVICES OFFICE					
GOVERNMENT OF GUAM					
P.O. BOX 2877					
AGUANA, GU 96910					
(671) 472-7230					
FTS 550-7230					
24-Hour Emergency Number: NONE					
DR. GEORGE BOUGHTON	*				
CHAIR					
Contact(s):					
DR. GEORGE BOUGHTON		*	*	*	
GUAM EPA					
D-107 HARMON PLAZA					
130 ROJAS STREET					
HARMON, GU 96911					
Contact(s):					
FRED M. CASTRO					*
671-646-8863					

STATE EMERGENCY RESPONSE COMMISSIONS (SERCS)

	Chair	302	304	311/312	313
HAWAII					
HAWAII STATE EMERGENCY RESPONSE COMMISSION					
HAWAII STATE DEPARTMENT OF HEALTH					
P.O. BOX 3378					
HONOLULU, HI 96801-9904					
(808) 586-0369					
24-Hour Emergency Number: (808) 226-3799, 734-2161					
JOHN C. LEWIN, M.D.	*				
CHAIRMAN					
Contact(s):					
BRUCE ANDERSON, VICECHAIR		*	*		
(808) 586-4424					
MARK INGOGLIA					
(808) 586-4249					
HAWAII STATE EMERGENCY RESPONSE COMMISSION					
HAWAII STATE DEPARTMENT OF HEALTH					
P.O. BOX 3378					
HONOLULU, HI 96801-9904					
(808) 586-4353					
Contact(s):					
RHONDA GOYKE				*	
HAWAII STATE EMERGENCY RESPONSE COMMISSION					
HAWAII DEPARTMENT OF HEALTH					
5 WATERFRONT PLAZA, SUITE 250					
500 ALAMONA BOULEVARD					
HONOLULU, HI 96813					
(808) 586-4249					
Contact(s):					
STEVE ARMAN					*
IDAHO					
IDAHO EMERGENCY RESPONSE COMMISSION					
1410 N. HILTON					
BOISE, ID 83706					
(208) 334-5888					
24-Hour Emergency Number: (208) 327-7422					
JACK PETERSON	*				
CHAIRMAN					
Contact(s):					
MARGARET BALLARD		*	*	*	*
(208) 334-5888					

STATE EMERGENCY RESPONSE COMMISSIONS (SERCS)

	Chair	302	304	311/312	313
ILLINOIS					
STATE EMERGENCY RESPONSE COMMISSION					
C/O ILLINOIS EMERGENCY MANAGEMENT AGENCY					
ATTN: HAZMAT SECTION					
110 EAST ADAMS STREET					
SPRINGFIELD, IL 62706					
(217) 782-4694					
24-Hour Emergency Number: (217) 782-7860					
RON STEPHENS	*				
CHAIRMAN					
Contact(s):					
ORAN ROBINSON		*	*	*	
(217) 782-4694					
OFFICE OF CHEMICAL SAFETY					
ILLINOIS EPA					
P.O. BOX 19276					
2200 CHURCHHILL					
SPRINGFIELD, IL 62794-9276					
(217) 785-0830					
EMERGENCY PLANNING UNIT					
Contact(s):					
JOE GOODNER					*
INDIANA					
INDIANA EMERGENCY RESPONSE COMMISSION					
5500 WEST BRADBURY AVENUE					
INDIANAPOLIS, IN 46241					
(317) 243-5176					
24-Hour Emergency Number: (317) 241-4336					
KATHY PROSSER	*				
CO-CHAIR					
Contact(s):					
SKIP POWERS		*	*	*	*
(317) 243-5176					

STATE EMERGENCY RESPONSE COMMISSIONS (SERCS)

	Chair	302	304	311/312	313
INDIANA EMERGENCY RESPONSE COMMISSION					
5500 WEST BRADBURY AVENUE					
INDIANAPOLIS, IN 46241					
(317) 243-5176					
24-Hour Emergency Number: (317) 241-4336					
JEROME M. HAUER	*				
CO-CHAIR					
Contact(s):					
SKIP POWERS		*	*	*	*
(317) 243-5176					

IOWA

	Chair	302	304	311/312	313
IOWA EMERGENCY RESPONSE COMMISSION					
C/O DIVISION OF LABOR					
1000 EAST GRAND AVENUE					
DES MOINES, IA 50319					
(515) 281-6175					
WALTER JOHNSON	*				
CHAIRMAN					
Contact(s):					
DON PEDDY				*	
IOWA DISASTER SERVICES					
HOOVER BUILDING, LEVEL A					
ROOM 29					
DES MOINES, IA 50319-0113					
(515) 281-3231					
Contact(s):					
ELLEN GORDON		*			
PAUL SADLER					
DEPARTMENT OF NATURAL RESOURCES					
WALLACE OFFICE BUILDING					
900 EAST GRAND AVENUE					
DES MOINES, IA 50319					
(515) 281-8852					
24-Hour Emergency Number: (515) 281-3231					
Contact(s):					
PETE HAMLIN			*		*

STATE EMERGENCY RESPONSE COMMISSIONS (SERCS)

KANSAS

	Chair	302	304	311/312	313
KANSAS EMERGENCY RESPONSE COMMISSION					
MILLS BUILDING, SUITE 501					
109 S.W. 9TH STREET					
TOPEKA, KS 66612					
(913) 296-1690					
NANCY BROWN	*				
CHAIR					
Contact(s):					
KARL BIRNS		*		*	*
(913) 296-1690					
DIVISION OF EMERGENCY PREPAREDNESS					
P.O. BOX C300					
TOPEKA, KS 66601-0300					
(913) 266-1431					
24-Hour Emergency Number: (913) 296-3176					
SECTION 304 REPORTING					
Contact(s):					
FRANK MOUSSA			*		
(913) 266-1400					

KENTUCKY

	Chair	302	304	311/312	313
KENTUCKY EMERGENCY RESPONSE COMMISSION					
KENTUCKY DISASTER AND EMERGENCY SERVICES					
BOONE NATIONAL GUARD CENTER					
FRANKFORT, KY 40601-6168					
(502) 564-8660					
24-Hour Emergency Number: (502) 564-7815					
JAMES M. EVERETT	*				
CHAIRMAN					
Contact(s):					
LUCILLE ORLANDO		*	*	*	
(502) 564-5223					
KENTUCKY DEPARTMENT FOR ENVIRONMENTAL PROTECTION					
14 REILLY ROAD					
FRANKFORT, KY 40601-1132					
(502) 564-2150					
Contact(s):					
VALERIE HUDSON					*

STATE EMERGENCY RESPONSE COMMISSIONS (SERCS)

	Chair	302	304	311/312	313
LOUISIANA					
LOUISIANA EMERGENCY RESPONSE COMMISSION					
OFFICE OF STATE POLICE/TESS/RIGHT-TO-KNOW UNIT					
P.O. BOX 66614					
7901 INDEPENDENCE BOULEVARD					
BATON ROUGE, LA 70896					
24-Hour Emergency Number: (504) 925-6595					
LARRY MOUNCE	*				
CHAIRMAN					
Contact(s):					
ROBERT HAYES		*	*	*	
(504) 925-6113					
DEPARTMENT OF ENVIRONMENTAL QUALITY					
7290 BLUEBONNET BLVD					
BATON ROUGE, LA 70810					
(504) 765-0737					
Contact(s):					
JEANY ANDERSON-LABAR					*
MAINE					
STATE EMERGENCY RESPONSE COMMISSION					
STATION NUMBER 72					
AUGUSTA, ME 04333					
(207) 289-4080					
(800) 452-8735 (IN ME)					
24-Hour Emergency Number: (800) 452-4664					
DAVID D. BROWN	*				
CHAIRMAN					
Contact(s):					
RAYNA LEIBOWITZ		*	*	*	*
(207) 289-4080					

STATE EMERGENCY RESPONSE COMMISSIONS (SERCS)

	Chair	302	304	311/312	313
MARYLAND					
GOVERNOR'S EMERGENCY MANAGEMENT ADVISORY COUNCIL					
C/O MARYLAND EMERGENCY MANAGEMENT AGENCY					
2 SUDBROOK LANE, EAST					
PIKESVILLE, MD 21208					
(410) 486-4422					
BRIGADIER GENERAL DONALD B. BARSHAY	*				
CHAIRMAN					
Contact(s):					
JUNE L. SWEN					
(410) 486-4422					
SARA TITLE III REPORTING					
MARYLAND DEPARTMENT OF THE ENVIRONMENT					
TOXICS REGISTRY DIVISION					
2500 BROENING HIGHWAY					
BALTIMORE, MD 21224					
(410) 631-3800					
24-Hour Emergency Number: (410) 486-4422					
MARSHA WAYS					
Contact(s):					
PATRICIA WILLIAMS		*	*	*	*
MASSACHUSETTS					
MASSACHUSETTS EMERGENCY MANAGEMENT AGENCY					
P.O. BOX 1496					
400 WORCESTER RD.					
FRAMINGHAM, MA 01701-0317					
(508) 820-2000					
24-Hour Emergency Number: (508) 820-2000					
A. DAVID RODHAM	*				
DIRECTOR					
Contact(s):					
SGT. ROBERT SCOFIELD		*	*	*	*
(508) 820-2000					

STATE EMERGENCY RESPONSE COMMISSIONS (SERCS)

	Chair	302	304	311/312	313
MICHIGAN					
EMERGENCY PLANNING AND COMMUNITY RIGHT-TO-KNOW COMMISSION					
MICHIGAN DEPARTMENT OF NATURAL RESOURCES					
ENVIRONMENTAL RESPONSE DIVISION					
P.O. BOX 30028					
LANSING, MI 48909					
24-Hour Emergency Number: (800) 292-4706 IN MI, (517) 373-7660 OUTSIDE MI					
JACK BAILS	*				
CHAIRMAN					
Contact(s):					
KENT KANAGY		*	*	*	*
(517) 373-4802					
MINNESOTA					
MINNESOTA EMERGENCY RESPONSE COMMISSION					
175 BIGELOW BUILDING					
450 NORTH SYNDICATE					
ST. PAUL, MN 55104					
(612) 643-3000					
LEE TISCHLER	*				
DIRECTOR					
Contact(s):					
MINNESOTA EMERGENCY RESPONSE COMMISSION					
175 BIGELOW BUILDING					
450 NORTH SYNDICATE STREET					
ST. PAUL, MN 55104					
(612) 643-3000					
Contact(s):					
STEVE TOMLYANOVICH					*
(612) 643-3542					
MINNESOTA EMERGENCY RESPONSE COMMISSION					
175 BIGELOW BUILDING					
450 NORTH SYNDICATE					
ST. PAUL, MN 55104					
(612) 643-3004					
24-Hour Emergency Number: (800) 422-0798 IN MN; (612) 649-5451 OUT OF MN					
Contact(s):					
BOB DAHM		*	*	*	
(612) 643-3004					

STATE EMERGENCY RESPONSE COMMISSIONS (SERCS)

MISSISSIPPI	Chair	302	304	311/312	313
MISSISSIPPI EMERGENCY RESPONSE COMMISSION					
MISSISSIPPI EMERGENCY MANAGEMENT AGENCY					
P.O. BOX 4501					
JACKSON, MS 39296-4501					
(601) 960-9000					
24-Hour Emergency Number: (800) 222-6362					
J.E. MAHER	*				
CHAIRMAN					
Contact(s):					
JOHN DAVID BURNS		*	*	*	*
(601) 960-9000					

MISSOURI	Chair	302	304	311/312	313
MISSOURI EMERGENCY RESPONSE COMMISSION					
MISSOURI DEPARTMENT OF NATURAL RESOURCES					
P.O. BOX 3133					
JEFFERSON CITY, MO 65102					
(314) 526-3371					
24-Hour Emergency Number: (314) 634-2436					
DEAN MARTIN	*				
COORDINATOR					
Contact(s):					
DEAN MARTIN		*	*	*	*

MONTANA	Chair	302	304	311/312	313
MONTANA EMERGENCY RESPONSE COMMISSION					
DES/DEPARTMENT OF MILITARY AFFAIRS					
1100 N. LAST CHANCE GULCH					
HELENA, MT 59620					
(406) 444-3948					
24-Hour Emergency Number: (406) 444-6911					
WILLIAM K. GOOD, JR.	*				
TOM ELLERHOFF					
Contact(s):					
TOM ELLERHOFF					
(406) 444-3948					

	Chair	302	304	311/312	313
MONTANA EMERGENCY RESPONSE COMMISSION					
ENVIRONMENTAL SCIENCES DIVISION					
DEPARTMENT OF HEALTH & ENVIRONMENTAL SCIENCES					
COGSWELL BUILDING A-107					
HELENA, MT 59620					
(406) 444-3948					
24-Hour Emergency Number: (406) 444-6911					
TOM ELLERHOFF	*				
CO-CHAIR					
Contact(s):					
TOM ELLERHOFF		*	*	*	*
(406) 444-3948					

NEBRASKA

	Chair	302	304	311/312	313
NEBRASKA EMERGENCY RESPONSE COMMISSION					
STATE CIVIL DEFENSE AGENCY					
1300 MILITARY ROAD					
LINCOLN, NE 68508-1090					
(402) 471-3241					
24-Hour Emergency Number: (402) 471-4545					
MAJOR GENERAL STANLEY HENG	*				
CHAIRMAN					
Contact(s):					
DAVID CHAMBERS					
(402) 471-4230					
NEBRASKA EMERGENCY RESPONSE COMMISSION					
STATE OF NEBRASKA DEPARTMENT OF ENVIRONMENTAL CONTROL					
P.O. BOX 98922					
STATE HOUSE STATION					
LINCOLN, NE 68509-8922					
(402) 471-4208					
24-Hour Emergency Number: (402) 471-4545					
Contact(s):					
DAVE TEWES			*		
NEBRASKA EMERGENCY RESPONSE COMMISSION					
STATE OF NEBRASKA DEPARTMENT OF ENVIRONMENTAL CONTROL					
P.O. BOX 98922					
LINCOLN, NE 68509-8922					
(402) 471-4230					
24-Hour Emergency Number: (402) 471-4545					
Contact(s):					
JOHN STEINAUER		*		*	*

STATE EMERGENCY RESPONSE COMMISSIONS (SERCS)

	Chair	302	304	311/312	313
NEVADA					
STATE EMERGENCY RESPONSE COMMISSION					
555 WRIGHT WAY					
CARSON CITY, NV 89711-0900					
(702) 687-6973					
24-Hour Emergency Number: (702) 687-5300					
ROBERT ANDREWS	*				
COORDINATOR					
Contact(s):					
JAKE SNOW		*	*	*	
(702) 739-5150					
DIVISION OF EMERGENCY MANAGEMENT					
2525 SOUTH CARSON STREET					
CARSON CITY, NV 89710					
(702) 687-4240					
Contact(s):					
KATHY ESPARZA					*
NEW HAMPSHIRE					
NEW HAMPSHIRE OFFICE OF EMERGENCEY MANAGEMENT					
TITLE III PROGRAM					
STATE OFFICE PARK SOUTH					
107 PLEASANT STREET					
CONCORD, NH 03301					
(603) 271-2231					
24-Hour Emergency Number: (800) 346-4009, (603) 271-3636					
GEORGE L. IVERSON	*				
DIRECTOR					
Contact(s):					
LELAND KIMBALL		*	*	*	*
(603) 271-2231					

STATE EMERGENCY RESPONSE COMMISSIONS (SERCS)

NEW JERSEY

	Chair	302	304	311/312	313
NEW JERSEY DEPT. ENVIRONMENTAL PROTECTION AND ENERGY					
DIVISION OF BUREAU OF HAZARDOUS SUBSTANCES INFORMATION					
ENVIRONMENTAL SAFETY, HEALTH AND ANALYTICAL PROGRAMS					
CN-405					
TRENTON, NJ 08625					
(609) 984-3219					
24-Hour Emergency Number: (609) 292-7172; (292-1075 DAYTIME)					
SHIRLEE CHIFFMAN	*				
CHIEF					
Contact(s):					
STAN DELIKAT		*	*	*	*
(609) 292-6714					

NEW MEXICO

	Chair	302	304	311/312	313
NEW MEXICO EMERGENCY RESPONSE COMMISSION					
NEW MEXICO DEPARTMENT OF PUBLIC SAFETY					
P.O. BOX 1628					
SANTA FE, NM 87504-1628					
(505) 827-9223					
WILLIAM MOORE	*				
CHAIRMAN					
Contact(s):					
KEITH LOUGH					
(505) 827-9222					
NEW MEXICO EMERGENCY RESPONSE COMMISSION					
CHEMICAL SAFETY OFFICE					
EMERGENCY MANAGEMENT BUREAU					
P.O. BOX 1628					
SANTA FE, NM 87504-1628					
(505) 827-9223					
24-Hour Emergency Number: (505) 827-9126					
Contact(s):					
MAX JOHNSON		*	*	*	*

STATE EMERGENCY RESPONSE COMMISSIONS (SERCS)

	Chair	302	304	311/312	313
NEW YORK					
STATE EMERGENCY RESPONSE COMMISSION STATE EMERGENCY MANAGEMENT OFFICE BUILDING 22 STATE CAMPUS ALBANY, NY 12226-5000 (518) 457-2222					
THOMAS JORLING CHAIRMAN	*				
Contact(s): ANTHONY GERMANO, Dir. DONALD DEVITO, Assc. Dir.					
NEW YORK EMERGENCY RESPONSE COMMISSION C/O STATE DEPARTMENT OF ENVIRONMENTAL CONSERVATION BUREAU OF SPILL PREVENTION AND RESPONSE 50 WOLF ROAD/ROOM 340 ALBANY, NY 12233-3510 (518) 457-4107 24-Hour Emergency Number: 800 457-7362 (in NY); 518-457-7362 (out NY) Contact(s):					
WILLIAM MINER		*	*	*	*
NORTH CAROLINA					
NORTH CAROLINA EMERGENCY RESPONSE COMMISSION NORTH CAROLINA DIVISION OF EMERGENCY MANAGEMENT 116 WEST JONES STREET RALEIGH, NC 27603-1335 (919) 733-3844 IN NC: (800)451-1403 GENERAL INFORMATION ONLY 24-Hour Emergency Number: (919) 733-3867					
JOSEPH F. MYERS CHAIRMAN	*				
Contact(s): EMILY KILPATRICK (919) 733-3865		*	*	*	*

STATE EMERGENCY RESPONSE COMMISSIONS (SERCS)

	Chair	302	304	311/312	313
NORTH DAKOTA					
NORTH DAKOTA STATE DIVISION OF EMERGENCY MANAGEMENT					
P.O. BOX 5511					
BISMARK, ND 58502-5511					
(701) 224-2111					
24-Hour Emergency Number: (800) 472-2121 OUTSIDE ND: (701) 224-2121					
BOB JOHNSTON	*				
Contact(s):					
RONALD AFFELDT		*		*	*
STATE RADIO COMMUNICATIONS DEPARTMENT					
P.O. BOX 5511					
BISMARK, ND 58502					
(701) 224-2121					
Contact(s):					
LYLE GALLAGHER			*		
DIRECTOR OF COMMUNICATIONS					
NORTHERN MARIANA ISLANDS					
OFFICE OF THE GOVERNOR					
CAPITOL HILL					
COMMONWEALTH OF NORTHERN MARIANA ISLANDS					
SAIPAN, CM 96950					
INTERNATIONAL NUMBER (670) 322-9529					
24-Hour Emergency Number: NONE					
FELIX A. SASAMOTO	*				
CIVIL DEFENSE COORDINATOR					
Contact(s):					
FELIX A. SASAMOTO		*	*	*	
DIVISION OF ENVIRONMENTAL QUALITY					
P.O. BOX 1304					
SAIPAN, CM 96950					
INTERNATIONAL NUMBER (670) 234-6984					
Contact(s):					
RUSSELL MEECHAM, III					*

STATE EMERGENCY RESPONSE COMMISSIONS (SERCS)

	Chair	302	304	311/312	313
OHIO					
OHIO EMERGENCY RESPONSE COMMISSION					
OHIO ENVIRONMENTAL PROTECTION AGENCY					
OFFICE OF EMERGENCY RESPONSE					
P.O. BOX 1049					
COLUMBUS, OH 43266-0149					
(614) 644-2260					
GRANT W. WILKSON	*				
Contact(s):					
GRANT W. WILKINSON					
OHIO EMERGENCY RESPONSE COMMISSION					
OHIO ENVIRONMENTAL PROTECTION AGENCY					
OFFICE OF EMERGENCY RESPONSE					
P.O. BOX 1049					
COLUMBUS, OH 43266-0149					
24-Hour Emergency Number: (800) 282-9378 IN OH, (614) 224-0946 OUTSIDE OH					
Contact(s):					
KEN SCHULTZ		*	*	*	
(614) 644-2260					
JEFF BEATTIE					
(614) 644-2260					
DIVISION OF AIR POLLUTION CONTROL					
1800 WATERMARK DRIVE					
COLUMBUS, OH 43215					
(614) 644-3604					
Contact(s):					
CINDY DEWULF					*
OKLAHOMA					
DEPARTMENT OF POLLUTION CONTROL					
P.O. BOX 53504					
OKLAHOMA CITY, OK 73152					
(405) 271-4468					
STEVE THOMPSON	*				
Contact(s):					
STEVE THOMPSON					
CHAIRMAN					

	Chair	302	304	311/312	313
OKLAHOMA DEPARTMENT OF HEALTH					
ENVIRONMENTAL HEALTH ADMINISTRATION - 0200					
1000 N.E. TENTH ST.					
OKLAHOMA CITY, OK 73117-1299					
(405) 271-8056					
LAWRENCE A. GALES					
Contact(s):					
LARRY GALES		*		*	*
DEPARTMENT OF POLLUTION CONTROL					
P.O. BOX 53504					
OKLAHOMA CITY, OK 73152					
(405) 271-8056					
24-Hour Emergency Number: (800) 522-0206 IN OK, (405) 271-4468 OUTSIDE OK					
Contact(s):					
LYNN MOSS			*		

OREGON

	Chair	302	304	311/312	313
OREGON EMERGENCY RESPONSE COMMISSION					
C/O STATE FIRE MARSHALL					
4760 PORTLAND ROAD, NE					
SALEM, OR 97305-1760					
(503) 378-3473 ext.347					
24-Hour Emergency Number: (800) 452-0311 IN OR, (503) 378-6377 OUTSIDE OR					
MARTHA PAGEL	*				
CHAIRPERSON					
Contact(s):					
DENNIS WALTHALL		*	*	*	*
(503) 378-3473 ext.231					

PENNSYLVANIA

	Chair	302	304	311/312	313
PENNSYLVANIA EMERGENCY MANAGEMENT COUNCIL					
SARA TITLE III OFFICER					
FEMA RESPONSE AND RECOVERY					
P.O. BOX 3321					
HARRISBURG, PA 17105-3321					
(717) 783-8150					
24-Hour Emergency Number: (717) 783-8150					
LT. GOV. MARK S. SINGEL	*				
CHAIRMAN					
Contact(s):					
RICHARD RODNEY		*	*		
(717) 783-8150					

STATE EMERGENCY RESPONSE COMMISSIONS (SERCS)

	Chair	302	304	311/312	313
PENNSYLVANIA EMERGENCY MANAGEMENT COUNCIL					
C/O BUREAU OF RIGHT-TO-KNOW					
RM 1503/LABOR AND INDUSTRY BUILDING					
7TH & FORSTER STREETS					
HARRISBURG, PA 17120					
(717) 783-2071					
Contact(s):					
LYNN SNEAD				*	*
PUERTO RICO					
PUERTO RICO EMERGENCY RESPONSE COMMISSION					
ENVIRONMENTAL QUALITY BOARD					
P.O. BOX 11488					
FERNADEZ JUNCOS STATION					
SANTURCE, PR 00910					
(809) 767-8181					
24-Hour Emergency Number: (809) 383-5130, 383-5131, 766-2823 (WORKING HOUR					
PEDRO MALDONADO	*				
CHAIR					
Contact(s):					
PEDRO MALDONADO		*	*	*	*
RHODE ISLAND					
RHODE ISLAND EMERGENCY RESPONSE COMMISSION					
STATE HOUSE ROOM 27					
PROVIDENCE, RI 02903					
(401) 277-3039					
24-Hour Emergency Number: (401) 274-7745					
JOSEPH CARNEVALE, JR.	*				
DIRECTOR					
Contact(s):					
JOHN AUCOTT		*	*		
(401) 277-3039					
RHODE ISLAND DEPARTMENT OF LABOR					
DIVISION OF OCCUPATIONAL SAFETY					
220 ELMWOOD AVENUE					
PROVIDENCE, RI 02907					
(401) 457-1847					
Contact(s):					
ANTHONY DICICCIO				*	

STATE EMERGENCY RESPONSE COMMISSIONS (SERCS)

	Chair	302	304	311/312	313
DEPARTMENT OF ENVIRONMENTAL MANAGEMENT					
DIVISION OF AIR AND HAZARDOUS MATERIALS					
291 PROMENADE STREET					
ATTN: TOXIC RELEASE INVENTORY					
PROVIDENCE, RI 02908-5767					
(401) 277-2808					
Contact(s):					
MARTHA DELANEY MULCAHEY					*
SOUTH CAROLINA					
SOUTH CAROLINA EMERGENCY RESPONSE COMMISSION					
DIVISION OF PUBLIC SAFETY PROGRAMS					
OFFICE OF THE GOVERNOR					
1205 PENDLETON STREET					
COLUMBIA, SC 29201					
(803) 734-0425					
STAN M. MCKINNEY	*				
CHAIR					
Contact(s):					
STAN M. MCKINNEY					
SC-DEPARTMENT OF HEALTH AND ENVIRONMENTAL CONTROL					
2600 BULL STREET					
COLUMBIA, SC 29201					
(803) 935-6336					
24-Hour Emergency Number: (803) 253-6488					
Contact(s):					
MICHAEL JURAS		*	*	*	*
SOUTH DAKOTA					
SOUTH DAKOTA EMERGENCY RESPONSE COMMISSION					
ENVIRONMENTAL AND NATURAL RESOURCES					
JOE FOSS BUILDING					
523 EAST CAPITOL					
PIERRE, SD 57501-3181					
(605) 773-3296					
24-Hour Emergency Number: (605) 773-3231					
LEE ANN SMITH	*				
TITLE III COORDINATOR					
Contact(s):					
LEE ANN SMITH		*	*	*	*

STATE EMERGENCY RESPONSE COMMISSIONS (SERCS)

TENNESSEE

	Chair	302	304	311/312	313
TENNESSEE EMERGENCY RESPONSE COUNCIL					
TENNESSEE EMERGENCY MANAGEMENT AGENCY					
3041 SIDCO DRIVE					
NASHVILLE, TN 37204					
(615) 741-0001					
24-Hour Emergency Number: (800) 262-3300 IN TN, (800) 258-3300 OUTSIDE TN					
MR. LACY SUITER	*				
Contact(s):					
BETTY EAVES		*	*	*	*
(615) 741-2986					

TEXAS

	Chair	302	304	311/312	313
TEXAS EMERGENCY RESPONSE COMMISSION					
DIVISION OF EMERGENCY MANAGEMENT					
P.O. BOX 4087					
AUSTIN, TX 78773-0001					
(512) 465-2138					
24-Hour Emergency Number: (512) 465-2000					
DAVID HAUN	*				
COORDINATOR					
Contact(s):					
REBECCA MULLEN			*		
(512) 465-2138					
TEXAS DEPARTMENT OF HEALTH					
OCCUPATIONAL SAFETY PROGRAM					
1100 WEST 49TH STREET					
AUSTIN, TX 78756					
(512) 459-1611					
IN TX, (800)-452-2791					
Contact(s):					
PAULA MCKINNEY		*		*	
OFFICE OF POLLUTION PREVENTION					
TEXAS WATER COMMISSION					
P.O. BOX 13087					
AUSTIN, TX 78711-3087					
(512) 463-4119					
DAVID BARKER					
SUPERVISOR					
Contact(s):					
BECKY KURKA					*

STATE EMERGENCY RESPONSE COMMISSIONS (SERCS)

	Chair	302	304	311/312	313
UTAH					
COMPREHENSIVE EMERGENCY MANAGEMENT					
STATE OFFICE BUILDING ROOM 110					
SALT LAKE CITY, UT 84158-0136					
24-Hour Emergency Number: (801) 538-3400					
LORAYNE FRANK	*				
DIRECTOR					
Contact(s):					
WAYNE DEUSNUP		*	*		
(801) 584-3400					
UTAH HAZARDOUS CHEMICAL EMERGENCY RESPONSE COMMISSION					
UTAH DIVISION OF ENVIRONMENTAL QUALITY					
1950 WEST NORTH TEMPLE					
SALT LAKE CITY, UT 84116					
(801) 536-4100					
Contact(s):					
NEIL TAYLOR				*	*
VERMONT					
VERMONT EMERGENCY RESPONSE COMMISSION					
Capt., ESSEX POLICE DEPT					
81 MAIN STREET					
ESSEX Jct., VT 05452					
(802) 828-8331					
ROBERT YANDOW	*				
Contact(s):					
ROBERT YANDOW					
VERMONT STATE DEPARTMENT OF HEALTH					
10 BALDWIN STREET					
MONTPELIER, VT 05602					
(802) 828-2886					
Contact(s):					
RAY MCCANDLESS		*		*	*
DEPARTMENT of PUBLIC SAFETY					
103 SOUTH MAIN STREET					
WATERBURY,, VT 05676					
(802) 244-8721					
Contact(s):					
GEORGE LOWE			*		
SECRETARY OF THE SERC BOARD					

STATE EMERGENCY RESPONSE COMMISSIONS (SERCS)

VIRGIN ISLANDS

	Chair	302	304	311/312	313
DEPARTMENT OF PLANNING AND NATURAL RESOURCES					
U.S. VIRGIN ISLANDS EMER. RESPONSE COMMISSION TITLE III					
NISKY CENTER, SUITE 231					
CHARLOTTE AMALIE					
ST. THOMAS, VI 00802					
(809) 774-3320 EXTENSION 101 OR 102					
24-Hour Emergency Number:					
ROY E. ADAMS	*				
COMMISSIONER					
Contact(s):					
ROY E. ADAMS		*	*	*	*

VIRGINIA

	Chair	302	304	311/312	313
VIRGINIA EMERGENCY RESPONSE COUNCIL					
C/O VIRGINIA DEPARTMENT OF WASTE MANAGEMENT					
JAMES MONROE BUILDING / 14TH FLOOR					
101 NORTH 14TH STREET					
RICHMOND, VA 23219					
(804) 225-2631					
CATHY HARRIS	*				
DIRECTOR					
Contact(s):					
CATHY HARRIS					
VIRGINIA EMERGENCY RESPONSE COUNCIL					
C/O VIRGINIA DEPARTMENT OF WASTE MANAGEMENT					
JAMES MONROE BUILDING / 14TH FLOOR					
101 NORTH 14TH STREET					
RICHMOND, VA 23219					
(804) 225-2513					
24-Hour Emergency Number: (800) 468-8892 IN VA, (804) 674-2400 OUTSIDE VA					
Contact(s):					
DAVID TIMBERLINE		*	*	*	
VIRGINIA EMERGENCY RESPONSE COUNCIL					
C/O VIRGINIA DEPARTMENT OF WASTE MANAGEMENT					
JAMES MONROE BUILDING / 14TH FLOOR					
101 NORTH 14TH STREET					
RICHMOND, VA 23219					
(804) 225-2581					
Contact(s):					
SHARON KENNEALLY-BAXTER					*

STATE EMERGENCY RESPONSE COMMISSIONS (SERCS)

	Chair	302	304	311/312	313
WASHINGTON					
WASHINGTON EMERGENCY RESPONSE COMMISSION					
DEPARTMENT OF COMMUNITY DEVELOPMENT					
9TH AND COLUMBIA BUILDING					
OLYMPIA, WA 98504					
CHUCK CLARKE	*				
CHAIR					
Contact(s):					
CHUCK CLARKE					
DEPARTMENT OF ECOLOGY					
COMMUNITY RIGHT-TO-KNOW UNIT					
P.O. BOX 47659					
OLYMPIA,, WA 98504-7659					
(206) 438-7252					
24-Hour Emergency Number: (800) 258-5990 IN AND OUTSIDE WA					
Contact(s):					
IDELL HANSEN		*	*	*	*
WEST VIRGINIA					
WEST VIRGINIA EMERGENCY RESPONSE COMMISSION					
WEST VIRGINIA OFFICE OF EMERGENCY SERVICES					
MAIN CAPITAL BUILDING 1, RM. EB-80					
CHARLESTON, WV 25305					
(304) 558-5380					
24-Hour Emergency Number: (304) 558-5380					
CARL L. BRADFORD	*				
DIRECTOR					
Contact(s):					
CARL L. BRADFORD		*	*	*	*
WISCONSIN					
WISCONSIN EMERGENCY RESPONSE BOARD					
DIVISION OF EMERGENCY GOVERNMENT					
P.O. BOX 7865					
MADISON, WI 53707					
(608) 266-3232					
24-Hour Emergency Number: (608) 266-3232					
ROBERT THOMPSON	*				
DIRECTOR					
Contact(s):					
CHRIS BACON		*	*	*	
(608) 266-1899					

STATE EMERGENCY RESPONSE COMMISSIONS (SERCS)

	Chair	302	304	311/312	313
DEPARTMENT OF NATURAL RESOURCES					
P.O. BOX 7921					
MADISON, WI 53707					
(608) 266-9255					
Contact(s):					
RUSS DUNST					*

WYOMING

	Chair	302	304	311/312	313
WYOMING EMERGENCY RESPONSE COMMISSION					
WYOMING EMERGENCY MANAGEMENT AGENCY					
P.O. BOX 1709					
CHEYENNE, WY 82003					
(307) 777-7566					
24-Hour Emergency Number: (307) 777-4321					
NANCY KROIS	*				
Contact(s):					
BROOKE HEFNER		*	*	*	*
(307) 777-7566					

Emergency Response Planning Guide

Guidelines For The Preparation And Review Of The SARA Title III Portions Of The Minnesota All-Hazard Emergency Plans

SARA, TITLE III PLANNING ELEMENTS

"Sec. 303. Comprehensive Emergency Response Plans

(c) PLAN PROVISIONS.-Each emergency plan shall include (but is not limited to) each of the following:"

1(A) *"Identification of facilities subject to the requirements of this subtitle that are within the emergency planning district."*

Plans must identify and describe section 302 facilities. Local jurisdictions may, at their discretion, expand the scope of their planning to include additional facilities which store other hazardous substances.

REQUIRED

Information which must be provided for section 302 facilities includes the facility name, location (street address), and a list of the extremely hazardous substances (EHSs) located at each facility. When other facilities are included in the plan, information which must be provided for those facilities includes the facility name, location (street address), and a selected list of the hazardous substances (HSs) located at each facility.

RECOMMENDED

The use of maps to identify facility locations is the preferred method however, they may be used only to clarify facility locations and may not be used as a substitute for the required information. Each facility should be mapped separately. Whenever practical, a single "summary" map showing the location of all facilities in the planning district should be provided. **"Required" information must be correlated with the facility maps.**

SPECIAL NOTE - COLOR CODING

A standardized color coding system for facility mapping has been established. While the use of this coding is not required, planners are encouraged to adopt the system in order to facilitate the coordination of mapping across jurisdictional boundaries.

Section 302 facilities should be identified in RED.
Non-Section 302 facilities should be identified in GREEN.

SPECIAL NOTE - SYMBOL CODING

A standardized symbol coding system for facility mapping has been established. While the use of this coding is not required, planners are encouraged to adopt the system in order to facilitate the coordination of mapping across jurisdictional boundaries.

Section 302 facilities should be identified by a STAR (ASTERISK). *
Non-section 302 facilities should be identified by a SQUARE. ☐

SPECIAL NOTE - MAPPING

When maps are used, local jurisdictions may elect to map only Section 302 facilities, or all facilities included in the plan.

SPECIAL NOTE - EHS LIST

The extremely hazardous substances which are required to be listed are only those substance located at the facility in quantities at or above the threshold planning quanitity.

1(B) *"identification of routes likely to be used for the transportation of substances on the list of extremely hazardous substances referred to in section 302(a)."*

The extent to which transportation routes are identified in the plan depends on the scope of the plan. For example, county maps may not provide sufficient detail of cities to permit the identification of secondary transportation routes within municipal boundaries, while city maps should permit identification of primary and secondary routes leading to and from individual facilities.

<u>REQUIRED</u>

Descriptions of primary and secondary routes must be provided. Information which must be provided includes the name/numbers of streets, roads, highways, pipelines, commercially navigable waterways, airports, and railroad lines used as primary and secondary transportation routes.

<u>RECOMMENDED</u>

The use of maps to identify primary and secondary transportation routes is the preferred method.

<u>SPECIAL NOTE - COLOR CODING</u>

A standardized color coding system for transportation mapping has been established. While the use of this coding is not required, planners are encouraged to adopt the system in order to facilitate the coordination of mapping across jurisdictional boundaries.

> Primary transportation routes should be highlighted in RED (PINK).
> Secondary transportation routes should be highlighted in GREEN.

<u>SPECIAL NOTE - SYMBOL CODING</u>

A standardized symbol coding system for transportation mapping has been established. While the use of this coding is not required, planners are encouraged to adopt the system in order to facilitate the coordination of mapping across jurisdictional boundaries.

> Primary transportation routes should be identified by a SOLID LINE. (__________)
> Secondary transportation routes should be identified by DOTS. (.............)

<u>SPECIAL NOTE - DEFINITIONS</u>

Routes are defined as streets, roads, highways, pipelines, commercially navigable waterways, airports, and railroad lines. Primary routes are defined as those routes contained within the planning district which are regularly, and most commonly used by motor vehicle and other forms of surface transportation. Secondary routes are defined as commonly used alternates to the primary routes. Secondary routes may also be those routes leading from a primary route to a specific facility, when such a facility is not located immediately adjacent to a primary route.

1(C) *"and identification of additional facilities contributing...additional risk due to their proximity to facilities subject to the requirements of this subtitle, such as...natural gas facilities."*

Local planners must use their discretion in determining whether a facility would contribute additional risk to the community by virtue of its proximity to a Section 302 facility. For example, is a non-section 302 facility which stores flammable liquids in close enough proximity to a section 302 facility such that an additional risk would be created by an incident at either location?

REQUIRED

Information which must be provided for facilities identified as contributing an additional risk includes the facility name, location (street address), and a selected list of the hazardous substances located at the facility.

RECOMMENDED

The name, work and home address, and 24 hour telephone number(s) of the facility's primary contact person, and at least one alternate contact, should be provided when available.

The use of maps to identify facilities contributing to additional risk is the preferred method however, they may be used only to clarify facility locations, and not as a substitute for the required information. **"Required" information must be correlated with the facility maps.**

SPECIAL NOTE - COLOR CODING

A standardized color coding system for facility mapping has been established. While the use of this coding is not required, planners are encouraged to adopt the system in order to facilitate the coordination of mapping across jurisdictional boundaries.

Facilities contributing to additional risk should be identified in BLUE.

SPECIAL NOTE - SYMBOL CODING

A standardized symbol coding system for facility mapping has been established. While the use of this coding is not required, planners are encouraged to adopt the system in order to facilitate the coordination of mapping across jurisdictional boundaries.

Facilities contributing to additional risk should be identified by a TRIANGLE. △

SPECIAL NOTE - FACILITY CONTACT PERSON

Facilities contributing to additional risk may be asked to provide the name, work and home address, and 24 hour telephone number(s) of a primary facility contact person, and at least one alternate contact, however, federal or state law does not require them to provide this information.

1(D) ***"and identification of additional facilities...subject to additional risk due to their proximity to facilities subject to the requirements of this subtitle, such as hospitals...".***

Examples of these facilities include hospitals, schools, nursing homes, shopping malls, day care centers, local government offices, fire stations, and law enforcement centers. Other types of facilities may also be identified by local planners.

REQUIRED

Information which must be provided for each facility includes the facility name, and location (street address).

RECOMMENDED

The name, work and home address, and 24 hour telephone number(s) of the facility's primary contact person, and at least one alternate should be provide when available.

The use of maps to identify facilities contributing to additional risk is the preferred method however, they may be used only to clarify facility locations, not as a substitute for the required information. **"Required" information must be correlated with the facility maps.**

Depending on the type of facility involved, it may also be beneficial to describe the occupancy. Occupancy may be described in terms of a specific group of people, such as senior citizens, seasonal usage, populations, time of day, stc.

SPECIAL NOTE - COLOR CODING

A standardized color coding system for facility mapping has been established. While the use of this coding is not required, planners are encouraged to adopt the system in order to facilitate the coordination of mapping across jurisdictional boundaries.

Facilities subject to risk should be identified in BLACK.

SPECIAL NOTE - SYMBOL CODING

A standardized symbol coding system for facility mapping has been established. While the use of this coding is not required, planners are encouraged to adopt the system in order to facilitate the coordination of mapping across jurisdictional boundaries.

Facilities subject to risk should be identified by a CIRCLE. O

Letters may be placed within the open circle to identify specific types of facilities. The following examples are offered as guidance:

A	Ambulance Station	L	Law Enforcement
C	Church	M	Shopping Mall
D	Day Care Center	N	Nursing Home
F	Fire Station	P	Park
G	Government Offices	R	Recreation Center
H	Hospital	S	School

As an alternative, numbers may be placed within the open circle to correlate with a numbered list of the facilities represented.

SPECIAL NOTE - FACILITY CONTACT PERSON

Facilities subject to additional risk may be asked to provide the name, work and home address, and 24 hour telephone number(s) of a primary facility contact person, and at least one alternate contact, however, federal or state law does not require them to provide this information.

2(A) ***"Methods and procedures to be followed by facility owners and operators...to respond to any release of such substances."***

This element relates to facility standard operating procedures which should be part of the facility emergency response plans, or emergency action plans, generally required of facilities by OSHA under SARA, Title I, or the US EPA under the Resource Conservation and Recovery Act (RCRA). The size and scope of such plans will vary depending upon the size of the facility and the amount and types of chemicals on the site. At the minimum, they should address the items discussed in planning element 4(A).

REQUIRED

The plan must identify section 302 facilities which have prepared an emergency response plan. Where facilities have not prepared such plans, the community plan must specify the facility(s) accidental release reporting requirements contained in SARA, Title III, Section 304, and as described under planning element 4(A).

RECOMMENDED

Planners may include a copy, or summary, of a facility's emergency response plan as a supplement to the community emergency plan. Such action should not be taken without the permission of the facility.

2(B) *"Methods and procedures to be followed by...local emergency and medical personnel to respond to any release of such substances."*

This element relates to the standard operating procedures (SOPs)/emergency response plans of community response agencies contained in the community emergency plan. Individual agency hazardous materials SOPs/emergency response plans, required by OSHA under SARA, Title I, to guide the response activities of emergency personnel should be developed by the various agencies.

REQUIRED

The plan must identify the community response agency(s) responsible for responding to a hazardous materials emergency, the level of that response, and the role of each agency.

RECOMMENDED

The location of each agency's hazardous materials response standard operating procedure/emergency response plan, and the title of the individual within each agency responsible for the development of such procedures, may be included in the plan.

Whenever practical, planners may elect to include copies of the response agency's standard operating procedures/emergency response plan as a supplement to the plan.

SPECIAL NOTE - LEVEL OF RESPONSE

The "level of response" information required under this planning element refers to the MN OSHA identified levels of hazardous materials response training, i.e. awareness, operations, technician, and specialist.

3(A) *"Designation of a community emergency coordinator...who shall make determinations necessary to implement the plan."*

The intent of this planning element is to identify the person or persons with authority to implement the community emergency plan in the event of a hazardous materials release. While more that one individual may hold such authority, at least during the initial stages of an emergency, a single individual must be designated as responsible for the overall implementation of the community emergency plan.

REQUIRED

The name, title, work and home address, and 24 hour telephone number(s) of the community emergency coordinator, and at least one alternate, must be provided.

3(B) *"Designation of...facility emergency coordinators, who shall make determinations necessary to implement the plan."*

The intent of this planning element is to not only provide the community with a contact person during a hazardous materials emergency, but to provide for ongoing contact and coordination during, and following, the planning process.

Section 302 facilities are required to provide planners with the name of the facility's emergency coordinator.

REQUIRED

The plan must include the name, work and home address, and 24 hour telephone number(s) of each section 302 facility emergency coordinator, and at least one alternate.

RECOMMENDED

When planners elect to include non-section 302 facilities under planning element 6.2 the name, work and home address, and 24 hour telephone number(s) of each facility's primary contact person, and at least one alternate, should be provided.

SPECIAL NOTE - FACILITY CONTACT PERSON

Non-section 302 facilities may be asked to provide the name, work and home address, and 24 hour telephone number(s) of a primary facility contact person, and at least one alternate contact, however, federal or state law does not require them to provide this information.

4(A) *"Procedures providing reliable, effective, and timely notification by the facility emergency coordinators...to persons designated in the emergency plan, and to the public, that a release has occurred (consistent with the emergency notification requirements of section 304)."*

Section 304 of Title III, the Minnesota Emergency Planning and Community Right-to-Know Act, and SERC procedures describes the notifications which section 302 facilities are required to make when an accidental release of an EHS occurs. In addition, some facilities may have specialized systems in place which also serve to warn the public of a hazardous materials release.

REQUIRED

The plan must specify the accidental release reporting requirements of SARA, Title III, Section 304; stating that section 302 facilities are required to contact local authorities (9-1-1), the State Response Center (DEM Duty Officer), and the National Response Center, whenever an accidental release occurs which is subject to the reporting requirements of section 304. These emergency notification telephone numbers, plus any others of local significance, must be identified in the plan.

When facilities have public warning systems in place, other than those activated by local authorities, the procedure by which they are activated and the method by which their activation is coordinated with the public systems, must be described.

RECOMMENDED

The plan may also state that a section 302 facility has completed its local public notification requirement when the facility emergency coordinator, or designee, makes initial contact with local authorities via 9-1-1.

4(B) *"Procedures providing reliable, effective and timely notification by...the community emergency coordinator to persons designated in the emergency plan, and to the public, that a release has occurred (consistent with the emergency notification requirements of section 304)."*

This planning element is intended to describe the public warning and notification system present in the community, the procedures by which it may be activated, and the sources of public information available to community residents.

REQUIRED

The plan must describe the public warning and notification system(s) available in the community. Information which must be provided includes the location and warning radius of local siren systems, call letters and frequencies of Emergency Broadcast System participating radio and television stations, identification of any cable access systems available for such purposes, and other means by which the public may be warned of the release of hazardous materials. The description must also identify the methods and procedures by which such systems may be activated.

All community key-contact personnel having an emergency response/operations (i.e. police chief, fire chief, etc.) function must be identified in the plan. Information to be provided for these individuals includes their name, title, work and home address, and 24 hour telephone number(s). The procedure normally used for alerting such personnel must also be specified.

RECOMMENDED

Sample emergency broadcast messages, which could be adapted for use during a hazardous materials emergency, may be included in the plan.

Planners may wish to describe the actions which may be taken by local authorities to warn the public following the release of a hazardous substance. Inclusion of a flow chart would serve to detail those actions.

The plan may also identify the public education measures by which community residents may be made aware of the public warning and notification systems in place in the community, and the personal actions which should be taken when warnings are issued.

5(A) *"Methods for determining the occurrence of a release,...".*

Methods for determining that a release has occurred may range from sophisticated technological/electronic detection and warning systems to simple observation by facility employees.

REQUIRED

Describe the systems, methods, and/or procedures used by each section 302 facility to determine that the release of an extremely hazardous substance has occurred.

RECOMMENDED

When non-section 302 facilities are included, the plan should also describe the systems, methods, and/or procedures used at these facilities for determining that the release of a hazardous substance has occurred.

5(B) *"Methods for determining...the area or population likely to be affected by such a release."*

Determining the area at risk or population affected involves the conduct of a hazard analysis. Methodology used to conduct the analysis must be consistently applied to all facilities within the jurisdiction, and may include the use of CAMEO (or similar systems), ARCHIE, EPA/FEMA guidance, or information provided by the facility.

It is highly recommended that facilities be involved in the preparation of the hazard analysis. At the minimum, it is recommended that the results of any hazard analysis be shared with the affected facility, and offered for comment by the facility emergency coordinator.

REQUIRED

The plan must identify the methodology used to conduct the hazard analysis, and include a description of the boundaries of the risk area surrounding each section 302 facility. Descriptions must include an estimate of the population located within the risk area. Population ranges should be used whenever the population of an at risk area fluctuates widely due to season, time of day, or other variables.

RECOMMENDED

The use of maps to identify the risk area surrounding each section 302 facility is the preferred method. A separate map should be used for each facility. When ever practical, a "summary" map showing the risk areas surrounding all section 302 facilities in the planning district should be provided. When maps are used, the risk area should be identified by drawing the appropriate radius around the facility.

6(A) ***"A description of emergency equipment and facilities in the community...., and an identification of the persons responsible for such equipment and facilities."***

The primary focus of this planning element is the local availability of response resources suitable for use during a hazardous materials incident.

REQUIRED

The plan must identify the various types of specialized hazardous materials response equipment locally available for use in responding to a hazardous materials incident. The location of such equipment, and the name, work and home address, and 24 telephone number(s) of contact persons with the authority to release the equipment for use must be provided.

RECOMMENDED

In addition to those resources under the direct control of local officials, reference may also be made to exiting mutual aid agreements which might provide additional resources, as well as state and federal resources, and privately owned resources which would be available during an emergency.

6(B) ***"A description of emergency equipment and facilities...at each facility in the community subject to the requirements of this subtitle, and an identification of the persons responsible for such equipment and facilities."***

Certain larger facilities have hazardous materials response resources on site that may be available for use during an emergency at the facility, or made available to public agencies during an emergency at another facility.

REQUIRED

The plan must include a description of the specialized hazardous materials response equipment available at each section 302 facility. The name, work and home address, and 24 hour telephone number(s) of facility personnel with the authority to release the equipment for use during an emergency must be provided.

RECOMMENDED

Facilities which have agreed to make their emergency response equipment available to the community for use during emergencies occurring at other locations should be identified in the plan. Copies of any written agreements specifying the details of such usage should also be included.

7. ***"Evacuation plans, including those for a precautionary evacuation and alternative traffic routes."***

This planning element is intended to deal with protective actions to be taken by the population at risk in the event of a hazardous materials release.

REQUIRED

Information which must be provided includes the names/numbers of streets, roads and highways to be used as primary and secondary evacuation routes by each population at risk identified in the hazard analysis conducted under planning element 5(B).

Each of the following evacuation considerations must also be addressed; assignment of responsibility for ordering an evacuation, procedures for initiating a "shelter in place" option, provision of congregate care for evacuees, assignment of responsibility for conducting and coordinating the evacuation (including notification, traffic control and security), provisions for the movement of special populations, determination of re-entry procedures, and the identification of shelter locations outside each risk area.

RECOMMENDED

Mapping is the preferred method of identifying evacuation routes and shelter locations. A separate map should be used for each risk area identified in the hazard analysis conducted under planning element 5(B). Whenever practical, a single "summary" map showing all primary and secondary evacuation routes in the community should be provided.

SPECIAL NOTE - COLOR CODING

A standardized color coding system for evacuation route mapping has been established. While the use of this coding is not required, planners are encouraged to adopt the system in order to facilitate the coordination of mapping across jurisdictional boundaries.

Primary evacuation routes should be highlighted in YELLOW.
Secondary evacuation routes should be highlighted in BLUE.

SPECIAL NOTE - SYMBOL CODING

A standardized symbol coding system for evacuation route mapping has been established. While the use of this coding is not required, planners are encouraged to adopt the system in order to facilitate the coordination of mapping across jurisdictional boundaries.

Primary evacuation routes should be identified by a HASH LINE. (------)
Secondary evacuation routes should be identified by a BROKEN LINE. (____..____)

8. ***"Training programs, including schedules for training of local emergency response and medical personnel."***

REQUIRED

Planning element 2(B) identifies the level of response of the various agencies covered by the plan. A statement must be included that indicates that agency personnel will be trained to that level. The plan must identify the location of schedules for providing such training.

9. ***"Methods and schedules for exercising the emergency plan."***

For purposes of Title III, the type (table top, functional, or full-scale), and the frequency of hazardous materials exercises, are at the discretion of local authorities. Title I and related regulations may require more frequent exercise activity.

REQUIRED

A copy of the jurisdiction's emergency management exercise plan must be provided.

RECOMMENDED

Planners should coordinate exercise programs with local emergency response agencies to prevent duplication of effort. **When such coordination results in a more frequent hazardous materials exercise schedule, a copy of that schedule may be used in place of the regular emergency management exercise plan.**

GENERAL PLANNING GUIDANCE

MAP COLOR CODE STANDARDS*

COLOR	IDENTIFIER
RED	Section 302 Facilities
GREEN	Non-section 302 Facilities
BLUE	Facilities Contributing to Additional Risk
BLACK	Facilities Subject to Risk
RED (PINK)	Primary Transportation Routes
GREEN	Secondary Transportation Routes
YELLOW	Primary Evacuation Routes
BLUE	Secondary Evacuation Routes

MAP SYMBOL CODING STANDARDS*

SYMBOL		IDENTIFIER
*	STAR (ASTERISK)	Section 302 Facilities
□	SQUARE	Non-section 302 Facilities
△	TRIANGLE	Facilities Contributing Additional Risk
○	CIRCLE	Facilities Subject to Risk
————	SOLID LINE	Primary Transportation Route(s)
....	DOTS	Secondary Transportation Route(s)
--------	HASH LINE	Primary Evacuation Route(s)
——..——..	BROKEN LINE	Secondary Evacuation Route(s)

Letters may be placed within the open circle to identify specific types of facilities. The following examples are offered as guidance:

A	Ambulance Station	L	Law Enforcement
C	Church	M	Shopping Mall
D	Day Care Center	N	Nursing Home
F	Fire Station	P	Park
G	Government Offices	R	Recreation Center
H	Hospital	S	School

As an alternative, numbers may be placed within the open circle to correlate with a numbered list of the facilities represented.

* - Use of this coding is not required however, planners are encouraged to adopt the system in order to facilitate the coordination of mapping across jurisdictional boundaries.

MAPPING

The use of maps to facilitate compliance with planning elements 1(A), 1(B), 1(C), 1(D), 5(B), and 7 does not require a separate map for each planning element. Whenever practical, planning element maps should be consolidated on a facility by facility basis, with a single "summary" map showing all the mapped planning elements should be provided.

COUNTY PLAN CONTENTS

County plans need not duplicate the planning of cities within those counties. However, those county plans must contain a complete list of all section 302 facilities located within the county, the city in which each facility is located, and identification of the city emergency plan which addresses the planning elements described in this guide.

County plans must also identify primary transportation routes throughout the county.

PLAN REVIEW: DOCUMENTS INCORPORATED BY REFERENCE

When planning requirements are satisfied by incorporating by reference information outside the plan, a sufficient description of that information must be given. A third person reading the plan should be provided with sufficient identifying information to locate the source that is incorporated by reference and the information within the source. Documents incorporated by reference should be themselves examined by the committee(s) as part of the plan review to make sure the incorporated documents satisfy the substantive requirements.

Consolidated List of Extremely Hazardous Substances and CERCLA Hazardous Substances

SARA TITLE III
CONSOLIDATED CHEMICAL LIST

This consolidated chemical list includes chemicals subject to reporting requirements under Title III of the Superfund Amendments and Reauthorization Act of 1986 (SARA)[1], also known as the Emergency Planning and Community Right-to-Know Act (EPCRA). It has been prepared to help firms handling chemicals determine whether they need to submit reports under sections 302, 304, or 313 of Title III and, for a specific chemical, what reports may need to be submitted. Separate lists are also provided of RCRA waste streams and unlisted hazardous wastes, and of radionuclides reportable under CERCLA. These lists should be used as a reference tool, not as a definitive source of compliance information. Compliance information is published in the Code of Federal Regulations, 40 CFR Parts 302, 355, and 372.

The chemicals on the consolidated list are ordered by Chemical Abstract Service (CAS) registry number. Categories of chemicals, which do not have CAS registry numbers, but which are cited under CERCLA and section 313, are placed at the end of the list. For reference purposes, the chemicals (with their CAS numbers) are ordered alphabetically following the CAS-order list. Long chemical names may have been truncated to facilitate printing of this list.

The list includes chemicals referenced under four federal statutory provisions, discussed below. More than one chemical name may be listed for one CAS number because the same chemical may appear on different lists under different names. For example, for CAS number 8001-35-2, the names toxaphene (from the section 313 list), camphechlor (from the section 302 list), and camphene, octachloro- (from the CERCLA list) all appear on this consolidated list. However, the chemicals listed under SARA Title III have many more synonyms than appear on this list.

(1) SARA Section 302 Extremely Hazardous Substances (EHSs)

The presence of EHSs in quantities in excess of the Threshold Planning Quantity (TPQ), requires certain emergency planning activities to be conducted. The extremely hazardous substances and their TPQs are listed in 40 CFR Part 355, Appendices A and B.

TPQ. The consolidated list presents the TPQ (in pounds) for section 302 chemicals in the column following the chemical name. For chemicals that are solids, there may be two TPQs given (e.g., 500/10,000). In these cases, the lower quantity applies for solids in powder form with particle size less than 100 microns, or if the substance is in solution or in molten form. Otherwise, the 10,000 pound TPQ applies.

EHS RQ. Releases of reportable quantities (RQ) of EHSs are subject to state and local reporting under section 304 of Title III. If a chemical listed under section 302 does not have a CERCLA RQ, a statutory RQ of one pound applies for section 304 reporting. The EHS RQ column lists the one-pound statutory RQ for EHSs not listed under CERCLA.

---

[1] **This consolidated list does not include all chemicals subject to the reporting requirements in sections 311 and 312 of SARA Title III. These hazardous chemicals, for which material safety data sheets (MSDS) must be developed under Occupational Safety and Health Act Hazard Communication Standards, are identified by broad criteria, rather than by enumeration. There are over 500,000 products that satisfy the criteria. See 40 CFR Part 370 for more information.**

(2) **CERCLA Hazardous Substances ("RQ chemicals")**

Releases of CERCLA hazardous substances, in quantities equal to or greater than their reportable quantity (RQ), are subject to reporting to the National Response Center under the Comprehensive Environmental Response, Compensation, and Liability Act of 1980 (CERCLA, or "Superfund"). Such releases are also subject to state and local reporting under section 304 of Title III. CERCLA hazardous substances, and their reportable quantities, are listed in 40 CFR Part 302, Table 302.4. On January 23, 1989, all section 302 chemicals not already listed under CERCLA were proposed for listing, and on August 30, 1989, adjusted RQs (not included in this document) were proposed for these chemicals. This document includes chemicals added to the CERCLA list because they are listed as hazardous air pollutants under section 112(b) of the Clean Air Act (CAA) of 1990. Radionuclides listed under CERCLA are provided in a separate list, with RQs in Curies.

RQ. The CERCLA RQ column in the consolidated list shows the RQs (in pounds) for chemicals that are CERCLA hazardous substances. An asterisk ("*") following the RQ indicates that no reporting of releases is required if the diameter of the pieces of the solid metal released is 100 micrometers (0.004 inches) or greater. Substances listed under CAA section 112(b) that have been added to the CERCLA list with statutory one-pound RQs are indicated by a plus sign ("+") following the RQ.

Note that the consolidated list does not include all CERCLA regulatory synonyms. See 40 CFR Part 302, Table 302.4 for a complete list.

(3) **SARA Section 313 Toxic Chemicals**

Emissions or releases of chemicals listed under section 313 must be reported annually as part of SARA Title III's community right-to-know provisions. The rule containing these chemicals was published on February 16, 1988 (53 FR 4500) (40 CFR Part 372).

Section 313. The notation "313" in the column for section 313 indicates that the chemical is subject to reporting under section 313 under the name listed. An "X" in this column indicates that the same chemical with the same CAS number appears on another list with a different chemical name.

Be aware that using or processing dissociable ammonium salts may produce ammonia in solution. Ammonia is reportable under section 313.

(4) **Chemical Categories**

The CERCLA and SARA section 313 lists include a number of chemical categories as well as specific chemicals. The chemicals on this consolidated list have not been systematiclly evaluated to determine whether they fall into any listed categories.

Some chemicals not specifically listed under CERCLA may be subject to CERCLA reporting as part of a category. For example, strychnine, sulfate (CAS number 60-41-3), listed under SARA section 302, is not on the CERCLA list, but may be subject to CERCLA reporting under the listing for strychnine and salts (CAS number 57-24-9), with an RQ of 10 pounds. Similarly, nicotine sulfate (CAS number 65-30-5) may be subject to CERCLA reporting under the listing for nicotine and salts (CAS number 54-11-5, RQ 100 pounds), and warfarin sodium (CAS number 129-06-6) may be subject to CERCLA reporting under the listing for warfarin and salts, concentration >0.3% (CAS number 81-88-9, RQ 100 pounds). The CERCLA list also includes a number of generic categories that have not been assigned RQs; chemicals falling into listed categories are considered CERCLA hazardous substances, but are not required to be reported under CERCLA unless an RQ has been assigned.

A number of chemical categories are subject to section 313 reporting. They appear at the end of the CAS number listing. Be aware that certain chemicals reportable under section 302 or CERCLA may belong to section 313 categories. For example, mercuric acetate (CAS number 1600-27-7), listed under section 302, is not specifically listed under section 313, but could be reported under section 313 as "Mercury Compounds" (no CAS number).

(5) RCRA Hazardous Wastes

The consolidated list includes specific chemicals from the P and U lists only (40 CFR 261.33). This listing is provided as an indicator that companies may already have data on a specific chemical that may be useful for Title III reporting. It is not intended to be a comprehensive list of RCRA P and U chemicals. RCRA hazardous wastes consisting of waste streams on the F and K lists, and wastes exhibiting the characteristics of ignitibility, corrosivity, reactivity, and EP toxicity, are provided in a separate list. The descriptions of the F and K waste streams have been abbreviated; see 40 CFR Part 302, Table 302.4 for complete descriptions.

RCRA Code. The letter-and-digit code in the RCRA Code column is the chemical's RCRA hazardous waste code.

Information Sources

For additional copies of this or other Title III documents, send requests to:

Section 313 Document Distribution Center
P.O. Box 12505
Cincinnati, OH 45212

Refer to document number EPA 560/4-92-012 for additional copies of this document.

A dBase version of this consolidated list is available on disk from:

National Technical Information Service (NTIS)
5285 Port Royal Road
Springfield, VA 22161
(703) 487-4600

Refer to PB89 158653.

Questions concerning changes to the list or other aspects of Title III may be submitted in writing to:

Emergency Planning and Community Right-to-Know Information Hotline
U.S. Environmental Protection Agency (OS-120)
401 M Street, SW
Washington, DC 20460

Alternatively, you may call the hotline at (800) 535-0202 between the hours of 8:30 AM and 7:30 PM Eastern Time.

CAS Number	Chemical Name	Sec. 302(EHS) TPQ	Section 304 EHS RQ	Section 304 CERCLA RQ	Sec 313	RCRA Code
50-00-0	Formaldehyde	500		100	313	U122
50-07-7	Mitomycin C	500/10,000		10		U010
50-14-6	Ergocalciferol	1,000/10,000	1			
50-18-0	Cyclophosphamide			10		U058
50-29-3	DDT			1		U061
50-32-8	Benzo[a]pyrene			1		U022
50-55-5	Reserpine			5,000		U200
51-21-8	Fluorouracil	500/10,000	1			
51-28-5	2,4-Dinitrophenol			10	313	P048
51-43-4	Epinephrine			1,000		P042
51-75-2	Nitrogen mustard	10	1		313	
51-75-2	Mechlorethamine	10	1		X	
51-79-6	Urethane			100	313	U238
51-79-6	Carbamic acid, ethyl ester			100	X	U238
51-79-6	Ethyl carbamate			100	X	U238
51-83-2	Carbachol chloride	500/10,000	1			
52-68-6	Trichlorfon			100	313	
52-85-7	Famphur			1,000		P097
53-70-3	Dibenz[a,h]anthracene			1		U063
53-96-3	2-Acetylaminofluorene			1	313	U005
54-11-5	Nicotine	100		100		P075
54-11-5	Pyridine, 3-(1-methyl-2-pyrrolidinyl)-,(S)	100		100		P075
54-11-5	Nicotine and salts			100		P075
54-62-6	Aminopterin	500/10,000	1			
55-18-5	N-Nitrosodiethylamine			1	313	U174
55-21-0	Benzamide				313	
55-63-0	Nitroglycerin			10	313	P081
55-91-4	Isofluorphate	100		100		P043
55-91-4	Diisopropylfluorophosphate	100		100		P043
56-04-2	Methylthiouracil			10		U164
56-23-5	Carbon tetrachloride			10	313	U211
56-25-7	Cantharidin	100/10,000	1			
56-38-2	Parathion	100		10	313	P089
56-49-5	3-Methylcholanthrene			10		U157
56-53-1	Diethylstilbestrol			1		U089
56-55-3	Benz[a]anthracene			10		U018
56-72-4	Coumaphos	100/10,000		10		
57-12-5	Cyanides (soluble salts and complexes)			10		P030
57-14-7	1,1-Dimethyl hydrazine	1,000		10	313	U098
57-14-7	Dimethylhydrazine	1,000		10	X	U098

CAS Number	Chemical Name	Sec. 302(EHS) TPQ	Section 304 EHS RQ	Section 304 CERCLA RQ	Sec 313	RCRA Code
57-14-7	Hydrazine, 1,1-dimethyl-	1,000		10	X	U098
57-24-9	Strychnine	100/10,000		10		P108
57-24-9	Strychnine, and salts			10		P108
57-47-6	Physostigmine	100/10,000	1			
57-57-8	beta-Propiolactone	500		1+	313	
57-64-7	Physostigmine, salicylate (1:1)	100/10,000	1			
57-74-9	Chlordane	1,000		1	313	U036
57-97-6	7,12-Dimethylbenz[a]anthracene			1		U094
58-36-6	Phenoxarsine, 10,10'-oxydi-	500/10,000	1			
58-89-9	Lindane	1,000/10,000		1	313	U129
58-89-9	Hexachlorocyclohexane (gamma isomer)	1,000/10,000		1	X	U129
58-90-2	2,3,4,6-Tetrachlorophenol			10		U212
59-50-7	p-Chloro-m-cresol			5,000		U039
59-88-1	Phenylhydrazine hydrochloride	1,000/10,000	1			
59-89-2	N-Nitrosomorpholine			1+	313	
60-00-4	Ethylenediamine-tetraacetic acid (EDTA)			5,000		
60-09-3	4-Aminoazobenzene				313	
60-11-7	4-Dimethylaminoazobenzene			10	313	U093
60-11-7	Dimethylaminoazobenzene			10	X	U093
60-29-7	Ethyl ether			100		U117
60-34-4	Methyl hydrazine	500		10	313	P068
60-35-5	Acetamide			1+	313	
60-41-3	Strychnine, sulfate	100/10,000	1			
60-51-5	Dimethoate	500/10,000		10		P044
60-57-1	Dieldrin			1		P037
61-82-5	Amitrole			10		U011
62-38-4	Phenylmercury acetate	500/10,000		100		P092
62-38-4	Phenylmercuric acetate	500/10,000		100		P092
62-44-2	Phenacetin			100		U187
62-50-0	Ethyl methanesulfonate			1		U119
62-53-3	Aniline	1,000		5,000	313	U012
62-55-5	Thioacetamide			10	313	U218
62-56-6	Thiourea			10	313	U219
62-73-7	Dichlorvos	1,000		10	313	
62-74-8	Sodium fluoroacetate	10/10,000		10		P058
62-74-8	Fluoroacetic acid, sodium salt	10/10,000		10		P058
62-75-9	N-Nitrosodimethylamine	1,000		10	313	P082
62-75-9	Nitrosodimethylamine	1,000		10	X	P082
62-75-9	Methanamine, N-methyl-N-nitroso-	1,000		10	X	P082
63-25-2	Carbaryl			100	313	

+ Listed as hazardous air pollutant under section 112(b) of the Clear Air Act; statutory RQ of one pound applies until RQs are adjusted.

CAS Number	Chemical Name	Sec. 302(EHS) TPQ	Section 304 EHS RQ	Section 304 CERCLA RQ	Sec 313	RCRA Code
64-00-6	Phenol, 3-(1-methylethyl)-, methylcarbamat	500/10,000	1			
64-18-6	Formic acid			5,000		U123
64-19-7	Acetic acid			5,000		
64-67-5	Diethyl sulfate			1+	313	
64-86-8	Colchicine	10/10,000	1			
65-30-5	Nicotine sulfate	100/10,000	1			
65-85-0	Benzoic acid			5,000		
66-75-1	Uracil mustard			10		U237
66-81-9	Cycloheximide	100/10,000	1			
67-56-1	Methanol			5,000	313	U154
67-63-0	Isopropyl alcohol (mfg-strong acid process				313	
67-64-1	Acetone			5,000	313	U002
67-66-3	Chloroform	10,000		10	313	U044
67-72-1	Hexachloroethane			100	313	U131
68-12-2	Dimethylformamide			1+		
68-76-8	Triaziquone				313	
70-25-7	Guanidine, N-methyl-N'-nitro-N-nitroso-			10		U163
70-30-4	Hexachlorophene			100		U132
70-69-9	Propiophenone, 4'-amino	100/10,000	1			
71-36-3	n-Butyl alcohol			5,000	313	U031
71-43-2	Benzene			10	313	U019
71-55-6	1,1,1-Trichloroethane			1,000	313	U226
71-55-6	Methyl chloroform			1,000	X	U226
71-63-6	Digitoxin	100/10,000	1			
72-20-8	Endrin	500/10,000		1		P051
72-43-5	Methoxychlor			1	313	U247
72-54-8	DDD			1		U060
72-55-9	DDE			1		
72-57-1	Trypan blue			10		U236
74-83-9	Bromomethane	1,000		1,000	313	U029
74-83-9	Methyl bromide	1,000		1,000	X	U029
74-85-1	Ethylene				313	
74-87-3	Chloromethane			100	313	U045
74-87-3	Methyl chloride			100	X	U045
74-88-4	Methyl iodide			100	313	U138
74-89-5	Monomethylamine			100		
74-90-8	Hydrogen cyanide	100		10	313	P063
74-90-8	Hydrocyanic acid	100		10	X	P063
74-93-1	Methyl mercaptan	500		100		U153
74-93-1	Thiomethanol	500		100		U153

+ Listed as hazardous air pollutant under section 112(b) of the Clear Air Act; statutory RQ of one pound applies until RQs are adjusted.

CAS Number	Chemical Name	Sec. 302(EHS) TPQ	Section 304 EHS RQ	Section 304 CERCLA RQ	Sec 313	RCRA Code
74-95-3	Methylene bromide			1,000	313	U068
75-00-3	Chloroethane			100	313	
75-00-3	Ethyl chloride			100	X	
75-01-4	Vinyl chloride			1	313	U043
75-04-7	Monoethylamine			100		
75-05-8	Acetonitrile			5,000	313	U003
75-07-0	Acetaldehyde			1,000	313	U001
75-09-2	Dichloromethane			1,000	313	U080
75-09-2	Methylene chloride			1,000	X	U080
75-15-0	Carbon disulfide	10,000		100	313	P022
75-20-7	Calcium carbide			10		
75-21-8	Ethylene oxide	1,000		10	313	U115
75-21-8	Oxirane	1,000		10	X	U115
75-25-2	Bromoform			100	313	U225
75-25-2	Tribromomethane			100	X	U225
75-27-4	Dichlorobromomethane			5,000	313	
75-34-3	1,1-Dichloroethane			1,000		U076
75-35-4	Vinylidene chloride			100	313	U078
75-35-4	1,1-Dichloroethylene			100	X	U078
75-36-5	Acetyl chloride			5,000		U006
75-44-5	Phosgene	10		10	313	P095
75-50-3	Trimethylamine			100		
75-55-8	Propyleneimine	10,000		1	313	P067
75-55-8	Aziridine, 2-methyl	10,000		1	X	P067
75-56-9	Propylene oxide	10,000		100	313	
75-60-5	Cacodylic acid			1		U136
75-63-8	Bromotrifluoromethane [Halon 1301]				313	
75-63-8	Halon 1301				X	
75-64-9	tert-Butylamine			1,000		
75-65-0	tert-Butyl alcohol				313	
75-69-4	Trichlorofluoromethane [CFC-11]			5,000	313	U121
75-69-4	CFC-11			5,000	X	U121
75-69-4	Trichloromonofluoromethane			5,000	X	U121
75-71-8	Dichlorodifluoromethane [CFC-12]			5,000	313	U075
75-71-8	CFC-12			5,000	X	U075
75-74-1	Tetramethyllead	100	1			
75-77-4	Trimethylchlorosilane	1,000	1			
75-78-5	Dimethyldichlorosilane	500	1			
75-79-6	Methyltrichlorosilane	500	1			
75-86-5	Acetone cyanohydrin	1,000		10		P069

CAS Number	Chemical Name	Sec. 302(EHS) TPQ	Section 304 EHS RQ	Section 304 CERCLA RQ	Sec 313	RCRA Code
75-87-6	Acetaldehyde, trichloro-			5,000		U034
75-99-0	2,2-Dichloropropionic acid			5,000		
76-01-7	Pentachloroethane			10		U184
76-02-8	Trichloroacetyl chloride	500	1			
76-13-1	Freon 113				313	
76-14-2	Dichlorotetrafluoroethane [CFC-114]				313	
76-14-2	CFC-114				X	
76-15-3	Monochloropentafluoroethane [CFC-115]				313	
76-15-3	CFC-115				X	
76-44-8	Heptachlor			1	313	P059
77-47-4	Hexachlorocyclopentadiene	100		10	313	U130
77-78-1	Dimethyl sulfate	500		100	313	U103
77-81-6	Tabun	10	1			
78-00-2	Tetraethyl lead	100		10		P110
78-34-2	Dioxathion	500	1			
78-53-5	Amiton	500	1			
78-59-1	Isophorone			5,000		
78-71-7	Oxetane, 3,3-bis(chloromethyl)-	500	1			
78-79-5	Isoprene			100		
78-81-9	iso-Butylamine			1,000		
78-82-0	Isobutyronitrile	1,000	1			
78-83-1	Isobutyl alcohol			5,000		U140
78-84-2	Isobutyraldehyde				313	
78-87-5	1,2-Dichloropropane			1,000	313	U083
78-87-5	Propane 1,2-dichloro-			1,000	X	U083
78-88-6	2,3-Dichloropropene			100	313	
78-92-2	sec-Butyl alcohol				313	
78-93-3	Methyl ethyl ketone			5,000	313	U159
78-93-3	Methyl ethyl ketone (MEK)			5,000	X	U159
78-94-4	Methyl vinyl ketone	10	1			
78-97-7	Lactonitrile	1,000	1			
78-99-9	1,1-Dichloropropane			1,000		
79-00-5	1,1,2-Trichloroethane			100	313	U227
79-01-6	Trichloroethylene			100	313	U228
79-06-1	Acrylamide	1,000/10,000		5,000	313	U007
79-09-4	Propionic acid			5,000		
79-10-7	Acrylic acid			5,000	313	U008
79-11-8	Chloroacetic acid	100/10,000		1+	313	
79-19-6	Thiosemicarbazide	100/10,000		100		P116
79-21-0	Peracetic acid	500	1		313	

+ Listed as hazardous air pollutant under section 112(b) of the Clean Air Act; statutory RQ of one pound applies until RQs are adjusted.

CAS Number	Chemical Name	Sec. 302(EHS) TPQ	Section 304 EHS RQ	Section 304 CERCLA RQ	Sec 313	RCRA Code
79-22-1	Methyl chloroformate	500		1,000		U156
79-31-2	iso-Butyric acid			5,000		
79-34-5	1,1,2,2-Tetrachloroethane			100	313	U209
79-44-7	Dimethylcarbamyl chloride			1	313	U097
79-46-9	2-Nitropropane			10	313	U171
80-05-7	4,4'-Isopropylidenediphenol				313	
80-15-9	Cumene hydroperoxide			10	313	U096
80-15-9	Hydroperoxide, 1-methyl-1-phenylethyl-			10	X	U096
80-62-6	Methyl methacrylate			1,000	313	U162
80-63-7	Methyl 2-chloroacrylate	500	1			
81-07-2	Saccharin (manufacturing)			100	313	U202
81-07-2	Saccharin and salts			100		U202
81-81-2	Warfarin	500/10,000		100		P001
81-81-2	Warfarin, & salts, conc.>0.3%			100		P001
81-88-9	C.I. Food Red 15				313	
82-28-0	1-Amino-2-methylanthraquinone				313	
82-66-6	Diphacinone	10/10,000	1			
82-68-8	Quintozene			100	313	U185
82-68-8	Pentachloronitrobenzene			100	X	U185
82-68-8	PCNB			100	X	U185
83-32-9	Acenaphthene			100		
84-66-2	Diethyl phthalate			1,000	313	U088
84-74-2	Dibutyl phthalate			10	313	U069
84-74-2	n-Butyl phthalate			10	X	U069
85-00-7	Diquat			1,000		
85-01-8	Phenanthrene			5,000		
85-44-9	Phthalic anhydride			5,000	313	U190
85-68-7	Butyl benzyl phthalate			100	313	
86-30-6	N-Nitrosodiphenylamine			100	313	
86-50-0	Azinphos-methyl	10/10,000		1		
86-50-0	Guthion	10/10,000		1		
86-73-7	Fluorene			5,000		
86-88-4	Antu	500/10,000		100		P072
86-88-4	Thiourea, 1-naphthalenyl-	500/10,000		100		P072
87-62-7	2,6-Xylidine				313	
87-65-0	2,6-Dichlorophenol			100		U082
87-68-3	Hexachloro-1,3-butadiene			1	313	U128
87-68-3	Hexachlorobutadiene			1	X	U128
87-86-5	Pentachlorophenol			10	313	U242
87-86-5	PCP			10	X	U242

CAS Number	Chemical Name	Sec. 302(EHS) TPQ	Section 304 EHS RQ	Section 304 CERCLA RQ	Sec 313	RCRA Code
88-05-1	Aniline, 2,4,6-trimethyl-	500	1			
88-06-2	2,4,6-Trichlorophenol			10	313	U230
88-72-2	o-Nitrotoluene			1,000		
88-75-5	2-Nitrophenol			100	313	
88-85-7	Dinoseb	100/10,000		1,000		P020
88-89-1	Picric acid				313	
90-04-0	o-Anisidine			1+	313	
90-43-7	2-Phenylphenol				313	
90-94-8	Michler's ketone				313	
91-08-7	Toluene-2,6-diisocyanate	100		100	313	
91-20-3	Naphthalene			100	313	U165
91-22-5	Quinoline			5,000	313	
91-58-7	2-Chloronaphthalene			5,000		U047
91-59-8	beta-Naphthylamine			10	313	U168
91-80-5	Methapyrilene			5,000		U155
91-94-1	3,3'-Dichlorobenzidine			1	313	U073
92-52-4	Biphenyl			1+	313	
92-67-1	4-Aminobiphenyl			1+	313	
92-87-5	Benzidine			1	313	U021
92-93-3	4-Nitrobiphenyl			1+	313	
93-72-1	Silvex (2,4,5-TP)			100		U233
93-76-5	2,4,5-T acid			1,000		U232
93-79-8	2,4,5-T esters			1,000		
94-11-1	2,4-D Esters			100		
94-36-0	Benzoyl peroxide				313	
94-58-6	Dihydrosafrole			10		U090
94-59-7	Safrole			100	313	U203
94-75-7	2,4-D			100	313	U240
94-75-7	2,4-D Acid			100	X	U240
94-75-7	2,4-D, salts and esters			100		U240
94-79-1	2,4-D Esters			100		
94-80-4	2,4-D Esters			100		
95-47-6	o-Xylene			1,000	313	U239
95-47-6	Benzene, o-dimethyl-			1,000	X	U239
95-48-7	o-Cresol	1,000/10,000		1,000	313	U052
95-50-1	1,2-Dichlorobenzene			100	313	U070
95-50-1	o-Dichlorobenzene			100	X	U070
95-53-4	o-Toluidine			100	313	U328
95-57-8	2-Chlorophenol			100		U048
95-63-6	1,2,4-Trimethylbenzene				313	

+ Listed as hazardous air pollutant under section 112(b) of the Clear Air Act; statutory RQ of one pound applies until RQs are adjusted.

CAS Number	Chemical Name	Sec. 302(EHS) TPQ	Section 304 EHS RQ	Section 304 CERCLA RQ	Sec 313	RCRA Code
95-80-7	2,4-Diaminotoluene			10	313	
95-94-3	1,2,4,5-Tetrachlorobenzene			5,000		U207
95-95-4	2,4,5-Trichlorophenol			10	313	U230
96-09-3	Styrene oxide			1+	313	
96-12-8	1,2-Dibromo-3-chloropropane			1	313	U066
96-12-8	DBCP			1	X	U066
96-33-3	Methyl acrylate				313	
96-45-7	Ethylene thiourea			10	313	U116
97-56-3	C.I. Solvent Yellow 3				313	
97-63-2	Ethyl methacrylate			1,000		U118
98-01-1	Furfural			5,000		U125
98-05-5	Benzenearsonic acid	10/10,000	1			
98-07-7	Benzoic trichloride	100		10	313	U023
98-07-7	Benzotrichloride	100		10	X	U023
98-09-9	Benzenesulfonyl chloride			100		U020
98-13-5	Trichlorophenylsilane	500	1			
98-16-8	Benzenamine, 3-(trifluoromethyl)-	500	1			
98-82-8	Cumene			5,000	313	U055
98-86-2	Acetophenone			5,000		U004
98-87-3	Benzal chloride	500		5,000	313	U017
98-88-4	Benzoyl chloride			1,000	313	
98-95-3	Nitrobenzene	10,000		1,000	313	U169
99-08-1	m-Nitrotoluene			1,000		
99-35-4	1,3,5-Trinitrobenzene			10		U234
99-55-8	5-Nitro-o-toluidine			100		U181
99-59-2	5-Nitro-o-anisidine				313	
99-65-0	m-Dinitrobenzene			100	313	
99-98-9	Dimethyl-p-phenylenediamine	10/10,000	1			
99-99-0	p-Nitrotoluene			1,000		
100-01-6	p-Nitroaniline			5,000		P077
100-02-7	4-Nitrophenol			100	313	U170
100-02-7	p-Nitrophenol			100	X	U170
100-14-1	Benzene, 1-(chloromethyl)-4-nitro-	500/10,000	1			
100-25-4	p-Dinitrobenzene			100	313	
100-41-4	Ethylbenzene			1,000	313	
100-42-5	Styrene			1,000	313	
100-44-7	Benzyl chloride	500		100	313	P028
100-47-0	Benzonitrile			5,000		
100-75-4	N-Nitrosopiperidine			10	313	U179
101-14-4	4,4'-Methylenebis(2-chloroaniline)			10	313	U158

+ Listed as hazardous air pollutant under section 112(b) of the Clear Air Act; statutory RQ of one pound applies until RQs are adjusted.

CAS Number	Chemical Name	Sec. 302(EHS) TPQ	Section 304 EHS RQ	CERCLA RQ	Sec 313	RCRA Code
101-14-4	MBOCA			10	X	U158
101-55-3	4-Bromophenyl phenyl ether			100		U030
101-61-1	4,4'-Methylenebis(N,N-dimethyl)benzenamine				313	
101-68-8	Methylenebis(phenylisocyanate)			1+	313	
101-68-8	MBI			1+	X	
101-77-9	4,4'-Methylenedianiline			1+	313	
101-80-4	4,4'-Diaminodiphenyl ether				313	
102-36-3	Isocyanic acid, 3,4-dichlorophenyl ester	500/10,000	1			
103-23-1	Bis(2-ethylhexyl) adipate				313	
103-85-5	Phenylthiourea	100/10,000		100		P093
104-94-9	p-Anisidine				313	
105-46-4	sec-Butyl acetate			5,000		
105-60-2	Caprolactam			1+		
105-67-9	2,4-Dimethylphenol			100	313	U101
106-42-3	p-Xylene			1,000	313	U239
106-42-3	Benzene, p-dimethyl-			1,000	X	U239
106-44-5	p-Cresol			1,000	313	U052
106-46-7	1,4-Dichlorobenzene			100	313	U072
106-47-8	p-Chloroaniline			1,000		P024
106-49-0	p-Toluidine			100		U353
106-50-3	p-Phenylenediamine			1+	313	
106-51-4	Quinone			10	313	U197
106-51-4	p-Benzoquinone			10	X	U197
106-88-7	1,2-Butylene oxide			1+	313	
106-89-8	Epichlorohydrin	1,000		100	313	U041
106-93-4	1,2-Dibromoethane			1	313	U067
106-93-4	Ethylene dibromide			1	X	U067
106-96-7	Propargyl bromide	10	1			
106-99-0	1,3-Butadiene			1+	313	
107-02-8	Acrolein	500		1	313	P003
107-05-1	Allyl chloride			1,000	313	
107-06-2	1,2-Dichloroethane			100	313	U077
107-06-2	Ethylene dichloride			100	X	U077
107-07-3	Chloroethanol	500	1			
107-10-8	n-Propylamine			5,000		U194
107-11-9	Allylamine	500	1			
107-12-0	Propionitrile	500		10		P101
107-12-0	Ethyl cyanide	500		10		P101
107-13-1	Acrylonitrile	10,000		100	313	U009
107-15-3	Ethylenediamine	10,000		5,000		

+ Listed as hazardous air pollutant under section 112(b) of the Clean Air Act; statutory RQ of one pound applies until RQs are adjusted.

CAS Number	Chemical Name	Sec. 302(EHS) TPQ	Section 304 EHS RQ	Section 304 CERCLA RQ	Sec 313	RCRA Code
107-16-4	Formaldehyde cyanohydrin	1,000	1			
107-18-6	Allyl alcohol	1,000		100	313	P005
107-19-7	Propargyl alcohol			1,000		P102
107-20-0	Chloroacetaldehyde			1,000		P023
107-21-1	Ethylene glycol			1+	313	
107-30-2	Chloromethyl methyl ether	100		10	313	U046
107-44-8	Sarin	10	1			
107-49-3	Tepp	100		10		P111
107-49-3	Tetraethyl pyrophosphate	100		10		P111
107-92-6	Butyric acid			5,000		
108-05-4	Vinyl acetate	1,000		5,000	313	
108-05-4	Vinyl acetate monomer	1,000		5,000	X	
108-10-1	Methyl isobutyl ketone			5,000	313	U161
108-23-6	Isopropyl chloroformate	1,000	1			
108-24-7	Acetic anhydride			5,000		
108-31-6	Maleic anhydride			5,000	313	U147
108-38-3	m-Xylene			1,000	313	U239
108-38-3	Benzene, m-dimethyl-			1,000	X	U239
108-39-4	m-Cresol			1,000	313	U052
108-46-3	Resorcinol			5,000		U201
108-60-1	Bis(2-chloro-1-methylethyl)ether			1,000	313	U027
108-60-1	Dichloroisopropyl ether			1,000	X	U027
108-88-3	Toluene			1,000	313	U220
108-90-7	Chlorobenzene			100	313	U037
108-91-8	Cyclohexylamine	10,000	1			
108-94-1	Cyclohexanone			5,000		U057
108-95-2	Phenol	500/10,000		1,000	313	U188
108-98-5	Thiophenol	500		100		P014
108-98-5	Benzenethiol	500		100		P014
109-06-8	2-Picoline			5,000		U191
109-61-5	Propyl chloroformate	500	1			
109-73-9	Butylamine			1,000		
109-77-3	Malononitrile	500/10,000		1,000		U149
109-86-4	2-Methoxyethanol				313	
109-89-7	Diethylamine			1,000		
109-99-9	Furan, tetrahydro-			1,000		U213
110-00-9	Furan	500		100		U124
110-16-7	Maleic acid			5,000		
110-17-8	Fumaric acid			5,000		
110-19-0	iso-Butyl acetate			5,000		

+ Listed as hazardous air pollutant under section 112(b) of the Clear Air Act; statutory RQ of one pound applies until RQs are adjusted.

CAS Number	Chemical Name	Sec. 302(EHS) TPQ	Section 304 EHS RQ	Section 304 CERCLA RQ	Sec 313	RCRA Code
110-54-3	Hexane			1+		
110-57-6	Trans-1,4-dichlorobutene	500	1			
110-75-8	2-Chloroethyl vinyl ether			1,000		U042
110-80-5	2-Ethoxyethanol			1,000	313	U359
110-80-5	Ethanol, 2-ethoxy-			1,000	X	U359
110-82-7	Cyclohexane			1,000	313	U056
110-86-1	Pyridine			1,000	313	U196
110-89-4	Piperidine	1,000	1			
111-42-2	Diethanolamine			1+	313	
111-44-4	Bis(2-chloroethyl) ether	10,000		10	313	U025
111-44-4	Dichloroethyl ether	10,000		10	X	U025
111-54-6	Ethylenebisdithiocarbamic acid, salts & es			5,000		U114
111-69-3	Adiponitrile	1,000	1			
111-91-1	Bis(2-chloroethoxy) methane			1,000		U024
114-26-1	Propoxur			1+	313	
115-02-6	Azaserine			1		U015
115-07-1	Propylene (Propene)				313	
115-21-9	Trichloroethylsilane	500	1			
115-26-4	Dimefox	500	1			
115-29-7	Endosulfan	10/10,000		1		P050
115-32-2	Dicofol			10	313	
115-90-2	Fensulfothion	500	1			
116-06-3	Aldicarb	100/10,000		1		P070
117-79-3	2-Aminoanthraquinone				313	
117-80-6	Dichlone			1		
117-81-7	Di(2-ethylhexyl) phthalate			100	313	U028
117-81-7	Bis(2-ethylhexyl)phthalate			100	X	U028
117-81-7	DEHP			100	X	U028
117-84-0	n-Dioctylphthalate			5,000	313	U107
117-84-0	Di-n-octyl phthalate			5,000	X	U107
118-74-1	Hexachlorobenzene			10	313	U127
119-38-0	Isopropylmethylpyrazolyl dimethylcarbamate	500	1			
119-90-4	3,3'-Dimethoxybenzidine			100	313	U091
119-93-7	3,3'-Dimethylbenzidine			10	313	U095
119-93-7	o-Tolidine			10	X	U095
120-12-7	Anthracene			5,000	313	
120-58-1	Isosafrole			100	313	U141
120-71-8	p-Cresidine				313	
120-80-9	Catechol			1+	313	
120-82-1	1,2,4-Trichlorobenzene			100	313	

+ Listed as hazardous air pollutant under section 112(b) of the Clear Air Act; statutory RQ of one pound applies until RQs are adjusted.

CAS Number	Chemical Name	Sec. 302(EHS) TPQ	Section 304 EHS RQ	Section 304 CERCLA RQ	Sec 313	RCRA Code
120-83-2	2,4-Dichlorophenol			100	313	U081
121-14-2	2,4-Dinitrotoluene			10	313	U105
121-21-1	Pyrethrins			1		
121-29-9	Pyrethrins			1		
121-44-8	Triethylamine			5,000		
121-69-7	N,N-Dimethylaniline			1+	313	
121-75-5	Malathion			100		
122-09-8	Benzeneethanamine, alpha,alpha-dimethyl-			5,000		P046
122-14-5	Fenitrothion	500	1			
122-66-7	1,2-Diphenylhydrazine			10	313	U109
122-66-7	Hydrazine, 1,2-diphenyl-			10	X	U109
122-66-7	Hydrazobenzene			10	X	U109
123-31-9	Hydroquinone	500/10,000		1+	313	
123-33-1	Maleic hydrazide			5,000		U148
123-38-6	Propionaldehyde			1+	313	
123-62-6	Propionic anhydride			5,000		
123-63-7	Paraldehyde			1,000		U182
123-72-8	Butyraldehyde				313	
123-73-9	Crotonaldehyde, (E)-	1,000		100		U053
123-86-4	Butyl acetate			5,000		
123-91-1	1,4-Dioxane			100	313	U108
123-92-2	iso-Amyl acetate			5,000		
124-04-9	Adipic acid			5,000		
124-40-3	Dimethylamine			1,000		U092
124-41-4	Sodium methylate			1,000		
124-48-1	Chlorodibromomethane			100		
124-65-2	Sodium cacodylate	100/10,000	1			
124-73-2	Dibromotetrafluoroethane [Halon 2402]				313	
124-73-2	Halon 2402				X	
124-87-8	Picrotoxin	500/10,000	1			
126-72-7	Tris(2,3-dibromopropyl) phosphate			10	313	U235
126-98-7	Methacrylonitrile	500		1,000		U152
126-99-8	Chloroprene			1+	313	
127-18-4	Tetrachloroethylene			100	313	U210
127-18-4	Perchloroethylene			100	X	U210
127-82-2	Zinc phenolsulfonate			5,000		
128-66-5	C.I. Vat Yellow 4				313	
129-00-0	Pyrene	1,000/10,000		5,000		
129-06-6	Warfarin sodium	100/10,000	1			
130-15-4	1,4-Naphthoquinone			5,000		U166

+ Listed as hazardous air pollutant under section 112(b) of the Clear Air Act; statutory RQ of one pound applies until RQs are adjusted.

CAS Number	Chemical Name	Sec. 302(EHS) TPQ	Section 304 EHS RQ	Section 304 CERCLA RQ	Sec 313	RCRA Code
131-11-3	Dimethyl phthalate			5,000	313	U102
131-74-8	Ammonium picrate			10		P009
131-89-5	2-Cyclohexyl-4,6-Dinitrophenol			100		P034
132-64-9	Dibenzofuran			1+	313	
133-06-2	Captan			10	313	
133-90-4	Chloramben			1+	313	
134-29-2	o-Anisidine hydrochloride				313	
134-32-7	alpha-Naphthylamine			100	313	U167
135-20-6	Cupferron				313	
137-26-8	Thiram			10		U244
139-13-9	Nitrilotriacetic acid				313	
139-65-1	4,4′-Thiodianiline				313	
140-29-4	Benzyl cyanide	500	1			
140-76-1	Pyridine, 2-methyl-5-vinyl-	500	1			
140-88-5	Ethyl acrylate			1,000	313	U113
141-32-2	Butyl acrylate				313	
141-66-2	Dicrotophos	100	1			
141-78-6	Ethyl acetate			5,000		U112
142-28-9	1,3-Dichloropropane			5,000		
142-71-2	Cupric acetate			100		
142-84-7	Dipropylamine			5,000		U110
143-33-9	Sodium cyanide (Na(CN))	100		10		P106
143-50-0	Kepone			1		U142
144-49-0	Fluoroacetic acid	10/10,000	1			
145-73-3	Endothall			1,000		P088
148-82-3	Melphalan			1		U150
149-74-6	Dichloromethylphenylsilane	1,000	1			
151-38-2	Methoxyethylmercuric acetate	500/10,000	1			
151-50-8	Potassium cyanide	100		10		P098
151-56-4	Ethyleneimine	500		1	313	P054
151-56-4	Aziridine	500		1	X	P054
152-16-9	Diphosphoramide, octamethyl-	100		100		P085
156-10-5	p-Nitrosodiphenylamine				313	
156-60-5	1,2-Dichloroethylene			1,000		U079
156-62-7	Calcium cyanamide			1+	313	
189-55-9	Dibenz[a,i]pyrene			10		U064
191-24-2	Benzo[ghi]perylene			5,000		
193-39-5	Indeno(1,2,3-cd)pyrene			100		U137
205-99-2	Benzo[b]fluoranthene			1		
206-44-0	Fluoranthene			100		U120

+ Listed as hazardous air pollutant under section 112(b) of the Clear Air Act; statutory RQ of one pound applies until RQs are adjusted.

CAS Number	Chemical Name	Sec. 302(EHS) TPQ	Section 304 EHS RQ	Section 304 CERCLA RQ	Sec 313	RCRA Code
207-08-9	Benzo(k)fluoranthene			5,000		
208-96-8	Acenaphthylene			5,000		
218-01-9	Chrysene			100		U050
225-51-4	Benz[c]acridine			100		U016
297-78-9	Isobenzan	100/10,000	1			
297-97-2	Thionazin	500		100		P040
297-97-2	O,O-Diethyl O-pyrazinyl phosphorothioate	500		100		P040
298-00-0	Parathion-methyl	100/10,000		100		P071
298-00-0	Methyl parathion	100/10,000		100		P071
298-02-2	Phorate	10		10		P094
298-04-4	Disulfoton	500		1		P039
300-62-9	Amphetamine	1,000	1			
300-76-5	Naled			10		
301-04-2	Lead acetate			5,000		U144
302-01-2	Hydrazine	1,000		1	313	U133
303-34-4	Lasiocarpine			10		U143
305-03-3	Chlorambucil			10		U035
309-00-2	Aldrin	500/10,000		1	313	P004
311-45-5	Diethyl-p-nitrophenyl phosphate			100		P041
315-18-4	Mexacarbate	500/10,000		1,000		
316-42-7	Emetine, dihydrochloride	1/10,000	1			
319-84-6	alpha-BHC			10		
319-85-7	beta-BHC			1		
319-86-8	delta-BHC			1		
327-98-0	Trichloronate	500	1			
329-71-5	2,5-Dinitrophenol			10		
330-54-1	Diuron			100		
333-41-5	Diazinon			1		
334-88-3	Diazomethane			1+	313	
353-42-4	Boron trifluoride compound with methyl eth	1,000	1			
353-50-4	Carbonic difluoride			1,000		U033
353-59-3	Bromochlorodifluoromethane [Halon 1211]				313	
353-59-3	Halon 1211				X	
357-57-3	Brucine			100		P018
359-06-8	Fluoroacetyl chloride	10	1			
371-62-0	Ethylene fluorohydrin	10	1			
379-79-3	Ergotamine tartrate	500/10,000	1			
460-19-5	Cyanogen			100		P031
463-58-1	Carbonyl sulfide			1+	313	
465-73-6	Isodrin	100/10,000		1		P060

+ Listed as hazardous air pollutant under section 112(b) of the Clear Air Act; statutory RQ of one pound applies until RQs are adjusted.

CAS Number	Chemical Name	Sec. 302(EHS) TPQ	Section 304 EHS RQ	Section 304 CERCLA RQ	Sec 313	RCRA Code
470-90-6	Chlorfenvinfos	500	1			
492-80-8	C.I. Solvent Yellow 34			100	313	U014
492-80-8	Auramine			100	X	U014
494-03-1	Chlornaphazine			100		U026
496-72-0	Diaminotoluene			10		
502-39-6	Methylmercuric dicyanamide	500/10,000	1			
504-24-5	Pyridine, 4-amino-	500/10,000		1,000		P008
504-24-5	4-Aminopyridine	500/10,000		1,000		P008
504-60-9	1,3-Pentadiene			100		U186
505-60-2	Mustard gas	500	1		313	
506-61-6	Potassium silver cyanide	500		1		P099
506-64-9	Silver cyanide			1		P104
506-68-3	Cyanogen bromide	500/10,000		1,000		U246
506-77-4	Cyanogen chloride			10		P033
506-78-5	Cyanogen iodide	1,000/10,000	1			
506-87-6	Ammonium carbonate			5,000		
506-96-7	Acetyl bromide			5,000		
509-14-8	Tetranitromethane	500		10		P112
510-15-6	Chlorobenzilate			10	313	U038
513-49-5	sec-Butylamine			1,000		
514-73-8	Dithiazanine iodide	500/10,000	1			
528-29-0	o-Dinitrobenzene			100	313	
532-27-4	2-Chloroacetophenone			1+	313	
534-07-6	Bis(chloromethyl) ketone	10/10,000	1			
534-52-1	Dinitrocresol	10/10,000		10		P047
534-52-1	4,6-Dinitro-o-cresol	10/10,000		10	313	P047
534-52-1	4,6-Dinitro-o-cresol and salts			10		P047
535-89-7	Crimidine	100/10,000	1			
538-07-8	Ethylbis(2-chloroethyl)amine	500	1			
540-59-0	1,2-Dichloroethylene				313	
540-73-8	Hydrazine, 1,2-dimethyl-			1		U099
540-84-1	2,2,4-Trimethylpentane			1+		
540-88-5	tert-Butyl acetate			5,000		
541-09-3	Uranyl acetate			100		
541-25-3	Lewisite	10	1			
541-41-3	Ethyl chloroformate				313	
541-53-7	Dithiobiuret	100/10,000		100		P049
541-73-1	1,3-Dichlorobenzene			100	313	U071
542-62-1	Barium cyanide			10		P013
542-75-6	1,3-Dichloropropylene			100	313	U084

+ Listed as hazardous air pollutant under section 112(b) of the Clear Air Act; statutory RQ of one pound applies until RQs are adjusted.

CAS Number	Chemical Name	Sec. 302(EHS) TPQ	Section 304 EHS RQ	Section 304 CERCLA RQ	Sec 313	RCRA Code
542-75-6	1,3-Dichloropropene			100	X	U084
542-76-7	Propionitrile, 3-chloro-	1,000		1,000		P027
542-76-7	3-Chloropropionitrile	1,000		1,000		P027
542-88-1	Bis(chloromethyl) ether	100		10	313	P016
542-88-1	Chloromethyl ether	100		10	X	P016
542-88-1	Dichloromethyl ether	100		10	X	P016
542-90-5	Ethylthiocyanate	10,000	1			
543-90-8	Cadmium acetate			10		
544-18-3	Cobaltous formate			1,000		
544-92-3	Copper cyanide			10		P029
554-84-7	m-Nitrophenol			100		
555-77-1	Tris(2-chloroethyl)amine	100	1			
556-61-6	Methyl isothiocyanate	500	1			
556-64-9	Methyl thiocyanate	10,000	1			
557-19-7	Nickel cyanide			10		P074
557-21-1	Zinc cyanide			10		P121
557-34-6	Zinc acetate			1,000		
557-41-5	Zinc formate			1,000		
558-25-8	Methanesulfonyl fluoride	1,000	1			
563-12-2	Ethion	1,000		10		
563-41-7	Semicarbazide hydrochloride	1,000/10,000	1			
563-68-8	Thallium(I) acetate			100		U214
569-64-2	C.I. Basic Green 4				313	
573-56-8	2,6-Dinitrophenol			10		
584-84-9	Toluene-2,4-diisocyanate	500		100	313	
591-08-2	1-Acetyl-2-thiourea			1,000		P002
592-01-8	Calcium cyanide			10		P021
592-04-1	Mercuric cyanide			1		
592-85-8	Mercuric thiocyanate			10		
592-87-0	Lead thiocyanate			100		
593-60-2	Vinyl bromide			1+	313	
594-42-3	Perchloromethylmercaptan	500		100		
594-42-3	Trichloromethanesulfenyl chloride	500		100		
597-64-8	Tetraethyltin	100	1			
598-31-2	Bromoacetone			1,000		P017
606-20-2	2,6-Dinitrotoluene			100	313	U106
608-93-5	Pentachlorobenzene			10		U183
609-19-8	3,4,5-Trichlorophenol			10		
610-39-9	3,4-Dinitrotoluene			10		
614-78-8	Thiourea, (2-methylphenyl)-	500/10,000	1			

+ Listed as hazardous air pollutant under section 112(b) of the Clear Air Act; statutory RQ of one pound applies until RQs are adjusted.

CAS Number	Chemical Name	Sec. 302(EHS) TPQ	Section 304 EHS RQ	Section 304 CERCLA RQ	Sec 313	RCRA Code
615-05-4	2,4-Diaminoanisole				313	
615-53-2	N-Nitroso-N-methylurethane			1		U178
621-64-7	N-Nitrosodi-n-propylamine			10	313	U111
621-64-7	Di-n-propylnitrosamine			10	X	U111
624-83-9	Methyl isocyanate	500		1	313	P064
625-16-1	tert-Amyl acetate			5,000		
626-38-0	sec-Amyl acetate			5,000		
627-11-2	Chloroethyl chloroformate	1,000	1			
628-63-7	Amyl acetate			5,000		
628-86-4	Mercury fulminate			10		P065
630-10-4	Selenourea			1,000		P103
630-20-6	Ethane, 1,1,1,2-tetrachloro-			100		U208
630-60-4	Ouabain	100/10,000	1			
631-61-8	Ammonium acetate			5,000		
636-21-5	o-Toluidine hydrochloride			100	313	U222
639-58-7	Triphenyltin chloride	500/10,000	1			
640-19-7	Fluoroacetamide	100/10,000		100		P057
644-64-4	Dimetilan	500/10,000	1			
675-14-9	Cyanuric fluoride	100	1			
676-97-1	Methyl phosphonic dichloride	100	1			
680-31-9	Hexamethylphosphoramide			1+	313	
684-93-5	N-Nitroso-N-methylurea			1	313	U177
692-42-2	Diethylarsine			1		P038
696-28-6	Phenyl dichloroarsine	500		1		P036
696-28-6	Dichlorophenylarsine	500		1		P036
732-11-6	Phosmet	10/10,000	1			
757-58-4	Hexaethyl tetraphosphate			100		P062
759-73-9	N-Nitroso-N-ethylurea			1	313	U176
760-93-0	Methacrylic anhydride	500	1			
764-41-0	2-Butene, 1,4-dichloro-			1		U074
765-34-4	Glycidylaldehyde			10		U126
786-19-6	Carbophenothion	500	1			
814-49-3	Diethyl chlorophosphate	500	1			
814-68-6	Acrylyl chloride	100	1			
815-82-7	Cupric tartrate			100		
822-06-0	Hexamethylene-1,6-diisocyanate			1+		
823-40-5	Diaminotoluene			10		
824-11-3	Trimethylolpropane phosphite	100/10,000	1			
842-07-9	C.I. Solvent Yellow 14				313	
900-95-8	Stannane, acetoxytriphenyl-	500/10,000	1			

+ Listed as hazardous air pollutant under section 112(b) of the Clear Air Act; statutory RQ of one pound applies until RQs are adjusted.

CAS Number	Chemical Name	Sec. 302(EHS) TPQ	Section 304 EHS RQ	Section 304 CERCLA RQ	Sec 313	RCRA Code
919-86-8	Demeton-S-methyl	500	1			
920-46-7	Methacryloyl chloride	100	1			
924-16-3	N-Nitrosodi-n-butylamine			10	313	U172
930-55-2	N-Nitrosopyrrolidine			1		U180
933-75-5	2,3,6-Trichlorophenol			10		
933-78-8	2,3,5-Trichlorophenol			10		
944-22-9	Fonofos	500	1			
947-02-4	Phosfolan	100/10,000	1			
950-10-7	Mephosfolan	500	1			
950-37-8	Methidathion	500/10,000	1			
959-98-8	alpha - Endosulfan			1		
961-11-5	Tetrachlorvinphos				313	
989-38-8	C.I. Basic Red 1				313	
991-42-4	Norbormide	100/10,000	1			
998-30-1	Triethoxysilane	500	1			
999-81-5	Chlormequat chloride	100/10,000	1			
1024-57-3	Heptachlor epoxide			1		
1031-07-8	Endosulfan sulfate			1		
1031-47-6	Triamiphos	500/10,000	1			
1066-30-4	Chromic acetate			1,000		
1066-33-7	Ammonium bicarbonate			5,000		
1066-45-1	Trimethyltin chloride	500/10,000	1			
1072-35-1	Lead stearate			5,000		
1111-78-0	Ammonium carbamate			5,000		
1116-54-7	N-Nitrosodiethanolamine			1		U173
1120-71-4	Propane sultone			10	313	U193
1120-71-4	1,3-Propane sultone			10	X	U193
1122-60-7	Nitrocyclohexane	500	1			
1124-33-0	Pyridine, 4-nitro-, 1-oxide	500/10,000	1			
1129-41-5	Metolcarb	100/10,000	1			
1163-19-5	Decabromodiphenyl oxide				313	
1185-57-5	Ferric ammonium citrate			1,000		
1194-65-6	Dichlobenil			100		
1300-71-6	Xylenol			1,000		
1303-28-2	Arsenic pentoxide	100/10,000		1		P011
1303-32-8	Arsenic disulfide			1		
1303-33-9	Arsenic trisulfide			1		
1306-19-0	Cadmium oxide	100/10,000	1			
1309-64-4	Antimony trioxide			1,000		
1310-58-3	Potassium hydroxide			1,000		

CAS Number	Chemical Name	Sec. 302(EHS) TPQ	Section 304 EHS RQ	Section 304 CERCLA RQ	Sec 313	RCRA Code
1310-73-2	Sodium hydroxide			1,000		
1313-27-5	Molybdenum trioxide				313	
1314-20-1	Thorium dioxide				313	
1314-32-5	Thallic oxide			100		P113
1314-56-3	Phosphorus pentoxide	10	1			
1314-62-1	Vanadium pentoxide	100/10,000		1,000		P120
1314-80-3	Sulfur phosphide			100		U189
1314-84-7	Zinc phosphide	500		100		P122
1314-84-7	Zinc phosphide (conc. <= 10%)	500		100		U249
1314-84-7	Zinc phosphide (conc. > 10%)	500		100		P122
1314-87-0	Lead sulfide			5,000		
1319-72-8	2,4,5-T amines			5,000		
1319-77-3	Cresol (mixed isomers)			1,000	313	U052
1320-18-9	2,4-D Esters			100		
1321-12-6	Nitrotoluene			1,000		
1327-52-2	Arsenic acid			1		P010
1327-53-3	Arsenous oxide	100/10,000		1		P012
1327-53-3	Arsenic trioxide	100/10,000		1		P012
1330-20-7	Xylene (mixed isomers)			1,000	313	U239
1332-07-6	Zinc borate			1,000		
1332-21-4	Asbestos (friable)			1	313	
1333-83-1	Sodium bifluoride			100		
1335-32-6	Lead subacetate			100		U146
1335-87-1	Hexachloronaphthalene				313	
1336-21-6	Ammonium hydroxide			1,000		
1336-36-3	Polychlorinated biphenyls			1	313	
1336-36-3	PCBs			1	X	
1338-23-4	Methyl ethyl ketone peroxide			10		U160
1338-24-5	Naphthenic acid			100		
1341-49-7	Ammonium bifluoride			100		
1344-28-1	Aluminum oxide (fibrous forms)				313	
1397-94-0	Antimycin A	1,000/10,000	1			
1420-07-1	Dinoterb	500/10,000	1			
1464-53-5	Diepoxybutane	500		10	313	U085
1464-53-5	2,2′-Bioxirane	500		10	X	U085
1558-25-4	Trichloro(chloromethyl)silane	100	1			
1563-66-2	Carbofuran	10/10,000		10		
1582-09-8	Trifluralin			1+	313	
1600-27-7	Mercuric acetate	500/10,000	1			
1615-80-1	Hydrazine, 1,2-diethyl-			10		U086

+ Listed as hazardous air pollutant under section 112(b) of the Clear Air Act; statutory RQ of one pound applies until RQs are adjusted.

CAS Number	Chemical Name	Sec. 302(EHS) TPQ	Section 304 EHS RQ	Section 304 CERCLA RQ	Sec 313	RCRA Code
1622-32-8	Ethanesulfonyl chloride, 2-chloro-	500	1			
1634-04-4	Methyl tert-butyl ether			1+	313	
1642-54-2	Diethylcarbamazine citrate	100/10,000	1			
1746-01-6	2,3,7,8-Tetrachlorodibenzo-p-dioxin (TCDD)			1		
1752-30-3	Acetone thiosemicarbazide	1,000/10,000	1			
1762-95-4	Ammonium thiocyanate			5,000		
1836-75-5	Nitrofen				313	
1863-63-4	Ammonium benzoate			5,000		
1888-71-7	Hexachloropropene			1,000		U243
1897-45-6	Chlorothalonil				313	
1910-42-5	Paraquat	10/10,000	1			
1918-00-9	Dicamba			1,000		
1928-38-7	2,4-D Esters			100		
1928-47-8	2,4,5-T esters			1,000		
1928-61-6	2,4-D Esters			100		
1929-73-3	2,4-D Esters			100		
1937-37-7	C.I. Direct Black 38				313	
1982-47-4	Chloroxuron	500/10,000	1			
2001-95-8	Valinomycin	1,000/10,000	1			
2008-46-0	2,4,5-T amines			5,000		
2032-65-7	Methiocarb	500/10,000		10		
2032-65-7	Mercaptodimethur	500/10,000		10		
2074-50-2	Paraquat methosulfate	10/10,000	1			
2097-19-0	Phenylsilatrane	100/10,000	1			
2104-64-5	EPN	100/10,000	1			
2164-17-2	Fluometuron				313	
2223-93-0	Cadmium stearate	1,000/10,000	1			
2231-57-4	Thiocarbazide	1,000/10,000	1			
2234-13-1	Octachloronaphthalene				313	
2238-07-5	Diglycidyl ether	1,000	1			
2275-18-5	Prothoate	100/10,000	1			
2303-16-4	Diallate			100	313	U062
2312-35-8	Propargite			10		
2497-07-6	Oxydisulfoton	500	1			
2524-03-0	Dimethyl phosphorochloridothioate	500	1			
2540-82-1	Formothion	100	1			
2545-59-7	2,4,5-T esters			1,000		
2570-26-5	Pentadecylamine	100/10,000	1			
2587-90-8	Phosphorothioic acid, O,O-dimethyl-5-(2-(m	500	1			
2602-46-2	C.I. Direct Blue 6				313	

+ Listed as hazardous air pollutant under section 112(b) of the Clear Air Act; statutory RQ of one pound applies until RQs are adjusted.

CAS Number	Chemical Name	Sec. 302(EHS) TPQ	Section 304 EHS RQ	Section 304 CERCLA RQ	Sec 313	RCRA Code
2631-37-0	Promecarb	500/10,000	1			
2636-26-2	Cyanophos	1,000	1			
2642-71-9	Azinphos-ethyl	100/10,000	1			
2665-30-7	Phosphonothioic acid, methyl-, O-(4-nitrop	500	1			
2703-13-1	Phosphonothioic acid, methyl-, O-ethyl O-(	500	1			
2757-18-8	Thallous malonate	100/10,000	1			
2763-96-4	Muscimol	500/10,000		1,000		P007
2763-96-4	5-(Aminomethyl)-3-isoxazolol	500/10,000		1,000		P007
2764-72-9	Diquat			1,000		
2778-04-3	Endothion	500/10,000	1			
2832-40-8	C.I. Disperse Yellow 3				313	
2921-88-2	Chlorpyrifos			1		
2944-67-4	Ferric ammonium oxalate			1,000		
2971-38-2	2,4-D Esters			100		
3012-65-5	Ammonium citrate, dibasic			5,000		
3037-72-7	Silane, (4-aminobutyl)diethoxymethyl-	1,000	1			
3118-97-6	C.I. Solvent Orange 7				313	
3164-29-2	Ammonium tartrate			5,000		
3165-93-3	4-Chloro-o-toluidine, hydrochloride			100		U049
3251-23-8	Cupric nitrate			100		
3254-63-5	Phosphoric acid, dimethyl 4-(methylthio) p	500	1			
3288-58-2	O,O-Diethyl S-methyl dithiophosphate			5,000		U087
3486-35-9	Zinc carbonate			1,000		
3547-04-4	DDE			1+		
3569-57-1	Sulfoxide, 3-chloropropyl octyl	500	1			
3615-21-2	Benzimidazole, 4,5-dichloro-2-(trifluorome	500/10,000	1			
3689-24-5	Sulfotep	500		100		P109
3689-24-5	Tetraethyldithiopyrophosphate	500		100		P109
3691-35-8	Chlorophacinone	100/10,000	1			
3734-97-2	Amiton oxalate	100/10,000	1			
3735-23-7	Methyl phenkapton	500	1			
3761-53-3	C.I. Food Red 5				313	
3813-14-7	2,4,5-T amines			5,000		
3878-19-1	Fuberidazole	100/10,000	1			
4044-65-9	Bitoscanate	500/10,000	1			
4098-71-9	Isophorone diisocyanate	100	1			
4104-14-7	Phosacetim	100/10,000	1			
4170-30-3	Crotonaldehyde	1,000		100		U053
4301-50-2	Fluenetil	100/10,000	1			
4418-66-0	Phenol, 2,2'-thiobis[4-chloro-6-methyl-	100/10,000	1			

+ Listed as hazardous air pollutant under section 112(b) of the Clear Air Act; statutory RQ of one pound applies until RQs are adjusted.

CAS Number	Chemical Name	Sec. 302(EHS) TPQ	Section 304 EHS RQ	Section 304 CERCLA RQ	Sec 313	RCRA Code
4549-40-0	N-Nitrosomethylvinylamine			10	313	P084
4680-78-8	C.I. Acid Green 3				313	
4835-11-4	Hexamethylenediamine, N,N'-dibutyl-	500	1			
5344-82-1	Thiourea, (2-chlorophenyl)-	100/10,000		100		P026
5836-29-3	Coumatetralyl	500/10,000	1			
5893-66-3	Cupric oxalate			100		
5972-73-6	Ammonium oxalate			5,000		
6009-70-7	Ammonium oxalate			5,000		
6369-96-6	2,4,5-T amines			5,000		
6369-97-7	2,4,5-T amines			5,000		
6484-52-2	Ammonium nitrate (solution)				313	
6533-73-9	Thallous carbonate	100/10,000		100		U215
6533-73-9	Thallium(I) carbonate	100/10,000		100		U215
6923-22-4	Monocrotophos	10/10,000	1			
7005-72-3	4-Chlorophenyl phenyl ether			5,000		
7421-93-4	Endrin aldehyde			1		
7428-48-0	Lead stearate			5,000		
7429-90-5	Aluminum (fume or dust)				313	
7439-92-1	Lead			1*	313	
7439-96-5	Manganese				313	
7439-97-6	Mercury			1	313	U151
7440-02-0	Nickel			100*	313	
7440-22-4	Silver			1,000*	313	
7440-23-5	Sodium			10		
7440-28-0	Thallium			1,000*	313	
7440-36-0	Antimony			5,000*	313	
7440-38-2	Arsenic			1*	313	
7440-39-3	Barium				313	
7440-41-7	Beryllium			10*	313	P015
7440-43-9	Cadmium			10*	313	
7440-47-3	Chromium			5,000*	313	
7440-48-4	Cobalt				313	
7440-50-8	Copper			5,000*	313	
7440-62-2	Vanadium (fume or dust)				313	
7440-66-6	Zinc (fume or dust)			1,000*	313	
7440-66-6	Zinc			1,000*		
7446-08-4	Selenium dioxide			10		
7446-09-5	Sulfur dioxide	500	1			
7446-11-9	Sulfur trioxide	100	1			
7446-14-2	Lead sulfate			100		

* No reporting of releases is required if the diameter of the pieces of the solid metal released is equal to or exceeds 100 micrometers (0.004 inches).

CAS Number	Chemical Name	Sec. 302(EHS) TPQ	Section 304 EHS RQ	Section 304 CERCLA RQ	Sec 313	RCRA Code
7446-18-6	Thallous sulfate	100/10,000		100		P115
7446-18-6	Thallium(I) sulfate	100/10,000		100		P115
7446-27-7	Lead phosphate			1		U145
7447-39-4	Cupric chloride			10		
7487-94-7	Mercuric chloride	500/10,000	1			
7488-56-4	Selenium sulfide			10		U205
7550-45-0	Titanium tetrachloride	100		1+	313	
7558-79-4	Sodium phosphate, dibasic			5,000		
7580-67-8	Lithium hydride	100	1			
7601-54-9	Sodium phosphate, tribasic			5,000		
7631-89-2	Sodium arsenate	1,000/10,000		1		
7631-90-5	Sodium bisulfite			5,000		
7632-00-0	Sodium nitrite			100		
7637-07-2	Boron trifluoride	500	1			
7645-25-2	Lead arsenate			1		
7646-85-7	Zinc chloride			1,000		
7647-01-0	Hydrochloric acid			5,000	313	
7647-01-0	Hydrogen chloride (gas only)	500		5,000	X	
7647-18-9	Antimony pentachloride			1,000		
7664-38-2	Phosphoric acid			5,000	313	
7664-39-3	Hydrogen fluoride	100		100	313	U134
7664-39-3	Hydrofluoric acid	100		100	X	U134
7664-41-7	Ammonia	500		100	313	
7664-93-9	Sulfuric acid	1,000		1,000	313	
7681-49-4	Sodium fluoride			1,000		
7681-52-9	Sodium hypochlorite			100		
7697-37-2	Nitric acid	1,000		1,000	313	
7699-45-8	Zinc bromide			1,000		
7705-08-0	Ferric chloride			1,000		
7718-54-9	Nickel chloride			100		
7719-12-2	Phosphorus trichloride	1,000		1,000		
7720-78-7	Ferrous sulfate			1,000		
7722-64-7	Potassium permanganate			100		
7722-84-1	Hydrogen peroxide (Conc.> 52%)	1,000	1			
7723-14-0	Phosphorus (yellow or white)	100		1	313	
7723-14-0	Phosphorus	100		1		
7726-95-6	Bromine	500	1			
7733-02-0	Zinc sulfate			1,000		
7738-94-5	Chromic acid			10		
7758-29-4	Sodium phosphate, tribasic			5,000		

+ Listed as hazardous air pollutant under section 112(b) of the Clear Air Act; statutory RQ of one pound applies until RQs are adjusted.

CAS Number	Chemical Name	Sec. 302(EHS) TPQ	Section 304 EHS RQ	Section 304 CERCLA RQ	Sec 313	RCRA Code
7758-94-3	Ferrous chloride			100		
7758-95-4	Lead chloride			100		
7758-98-7	Cupric sulfate			10		
7761-88-8	Silver nitrate			1		
7773-06-0	Ammonium sulfamate			5,000		
7775-11-3	Sodium chromate			10		
7778-39-4	Arsenic acid			1		P010
7778-44-1	Calcium arsenate	500/10,000		1		
7778-50-9	Potassium bichromate			10		
7778-54-3	Calcium hypochlorite			10		
7779-86-4	Zinc hydrosulfite			1,000		
7779-88-6	Zinc nitrate			1,000		
7782-41-4	Fluorine	500		10		P056
7782-49-2	Selenium			100*	313	
7782-50-5	Chlorine	100		10	313	
7782-63-0	Ferrous sulfate			1,000		
7782-82-3	Sodium selenite			100		
7782-86-7	Mercurous nitrate			10		
7783-00-8	Selenious acid	1,000/10,000		10		U204
7783-06-4	Hydrogen sulfide	500		100		U135
7783-07-5	Hydrogen selenide	10	1			
7783-20-2	Ammonium sulfate (solution)				313	
7783-35-9	Mercuric sulfate			10		
7783-46-2	Lead fluoride			100		
7783-49-5	Zinc fluoride			1,000		
7783-50-8	Ferric fluoride			100		
7783-56-4	Antimony trifluoride			1,000		
7783-60-0	Sulfur tetrafluoride	100	1			
7783-70-2	Antimony pentafluoride	500	1			
7783-80-4	Tellurium hexafluoride	100	1			
7784-34-1	Arsenous trichloride	500		1		
7784-40-9	Lead arsenate			1		
7784-41-0	Potassium arsenate			1		
7784-42-1	Arsine	100	1			
7784-46-5	Sodium arsenite	500/10,000		1		
7785-84-4	Sodium phosphate, tribasic			5,000		
7786-34-7	Mevinphos	500		10		
7786-81-4	Nickel sulfate			100		
7787-47-5	Beryllium chloride			1		
7787-49-7	Beryllium fluoride			1		

* No reporting of releases is required if the diameter of the pieces of the solid metal released is equal to or exceeds 100 micrometers (0.004 inches).

CAS Number	Chemical Name	Sec. 302(EHS) TPQ	Section 304 EHS RQ	Section 304 CERCLA RQ	Sec 313	RCRA Code
7787-55-5	Beryllium nitrate			1		
7788-98-9	Ammonium chromate			10		
7789-00-6	Potassium chromate			10		
7789-06-2	Strontium chromate			10		
7789-09-5	Ammonium bichromate			10		
7789-42-6	Cadmium bromide			10		
7789-43-7	Cobaltous bromide			1,000		
7789-61-9	Antimony tribromide			1,000		
7790-94-5	Chlorosulfonic acid			1,000		
7791-12-0	Thallous chloride	100/10,000		100		U216
7791-12-0	Thallium chloride TlCl	100/10,000		100		U216
7791-23-3	Selenium oxychloride	500	1			
7803-51-2	Phosphine	500		100		P096
7803-55-6	Ammonium vanadate			1,000		P119
8001-35-2	Toxaphene	500/10,000		1	313	P123
8001-35-2	Camphechlor	500/10,000		1	X	P123
8001-35-2	Camphene, octachloro-	500/10,000		1	X	P123
8001-58-9	Creosote			1	313	
8003-19-8	Dichloropropane - Dichloropropene (mixture			100		
8003-34-7	Pyrethrins			1		
8014-95-7	Sulfuric acid (fuming)			1,000		
8065-48-3	Demeton	500	1			
10022-70-5	Sodium hypochlorite			100		
10025-73-7	Chromic chloride	1/10,000	1			
10025-87-3	Phosphorus oxychloride	500		1,000		
10025-91-9	Antimony trichloride			1,000		
10026-11-6	Zirconium tetrachloride			5,000		
10026-13-8	Phosphorus pentachloride	500	1			
10028-15-6	Ozone	100	1			
10028-22-5	Ferric sulfate			1,000		
10031-59-1	Thallium sulfate	100/10,000		100		
10034-93-2	Hydrazine sulfate				313	
10039-32-4	Sodium phosphate, dibasic			5,000		
10043-01-3	Aluminum sulfate			5,000		
10045-89-3	Ferrous ammonium sulfate			1,000		
10045-94-0	Mercuric nitrate			10		
10049-04-4	Chlorine dioxide				313	
10049-05-5	Chromous chloride			1,000		
10099-74-8	Lead nitrate			100		
10101-53-8	Chromic sulfate			1,000		

CAS Number	Chemical Name	Sec. 302(EHS) TPQ	Section 304 EHS RQ	Section 304 CERCLA RQ	Sec 313	RCRA Code
10101-63-0	Lead iodide			100		
10101-89-0	Sodium phosphate, tribasic			5,000		
10102-06-4	Uranyl nitrate			100		
10102-18-8	Sodium selenite	100/10,000		100		
10102-20-2	Sodium tellurite	500/10,000	1			
10102-43-9	Nitric oxide	100		10		P076
10102-44-0	Nitrogen dioxide	100		10		P078
10102-45-1	Thallium(I) nitrate			100		U217
10102-48-4	Lead arsenate			1		
10108-64-2	Cadmium chloride			10		
10124-50-2	Potassium arsenite	500/10,000		1		
10124-56-8	Sodium phosphate, tribasic			5,000		
10140-65-5	Sodium phosphate, dibasic			5,000		
10140-87-1	Ethanol, 1,2-dichloro-, acetate	1,000	1			
10192-30-0	Ammonium bisulfite			5,000		
10196-04-0	Ammonium sulfite			5,000		
10210-68-1	Cobalt carbonyl	10/10,000	1			
10265-92-6	Methamidophos	100/10,000	1			
10294-34-5	Boron trichloride	500	1			
10311-84-9	Dialifor	100/10,000	1			
10361-89-4	Sodium phosphate, tribasic			5,000		
10380-29-7	Cupric sulfate, ammoniated			100		
10415-75-5	Mercurous nitrate			10		
10421-48-4	Ferric nitrate			1,000		
10476-95-6	Methacrolein diacetate	1,000	1			
10544-72-6	Nitrogen dioxide			10		
10588-01-9	Sodium bichromate			10		
11096-82-5	Aroclor 1260			1		
11097-69-1	Aroclor 1254			1		
11104-28-2	Aroclor 1221			1		
11115-74-5	Chromic acid			10		
11141-16-5	Aroclor 1232			1		
12002-03-8	Paris green	500/10,000		1		
12002-03-8	Cupric acetoarsenite	500/10,000		1		
12039-52-0	Selenious acid, dithallium(1+) salt			1,000		P114
12054-48-7	Nickel hydroxide			10		
12108-13-3	Manganese, tricarbonyl methylcyclopentadie	100	1			
12122-67-7	Zineb				313	
12125-01-8	Ammonium fluoride			100		
12125-02-9	Ammonium chloride			5,000		

CAS Number	Chemical Name	Sec. 302(EHS) TPQ	Section 304 EHS RQ	Section 304 CERCLA RQ	Sec 313	RCRA Code
12135-76-1	Ammonium sulfide			100		
12427-38-2	Maneb				313	
12672-29-6	Aroclor 1248			1		
12674-11-2	Aroclor 1016			1		
12771-08-3	Sulfur monochloride			1,000		
13071-79-9	Terbufos	100	1			
13171-21-6	Phosphamidon	100	1			
13194-48-4	Ethoprophos	1,000	1			
13410-01-0	Sodium selenate	100/10,000	1			
13450-90-3	Gallium trichloride	500/10,000	1			
13463-39-3	Nickel carbonyl	1		10		P073
13463-40-6	Iron, pentacarbonyl-	100	1			
13494-80-9	Tellurium	500/10,000	1			
13560-99-1	2,4,5-T salts			1,000		
13597-99-4	Beryllium nitrate			1		
13746-89-9	Zirconium nitrate			5,000		
13765-19-0	Calcium chromate			10		U032
13814-96-5	Lead fluoborate			100		
13826-83-0	Ammonium fluoborate			5,000		
13952-84-6	sec-Butylamine			1,000		
14017-41-5	Cobaltous sulfamate			1,000		
14167-18-1	Salcomine	500/10,000	1			
14216-75-2	Nickel nitrate			100		
14258-49-2	Ammonium oxalate			5,000		
14307-35-8	Lithium chromate			10		
14307-43-8	Ammonium tartrate			5,000		
14639-97-5	Zinc ammonium chloride			1,000		
14639-98-6	Zinc ammonium chloride			1,000		
14644-61-2	Zirconium sulfate			5,000		
15271-41-7	Bicyclo[2.2.1]heptane-2-carbonitrile, 5-ch	500/10,000	1			
15699-18-0	Nickel ammonium sulfate			100		
15739-80-7	Lead sulfate			100		
15950-66-0	2,3,4-Trichlorophenol			10		
16071-86-6	C.I. Direct Brown 95				313	
16543-55-8	N-Nitrosonornicotine				313	
16721-80-5	Sodium hydrosulfide			5,000		
16752-77-5	Methomyl	500/10,000		100		P066
16752-77-5	Ethanimidothioic acid, N-[[methylamino)car	500/10,000		100		P066
16871-71-9	Zinc silicofluoride			5,000		
16919-19-0	Ammonium silicofluoride			1,000		

CAS Number	Chemical Name	Sec. 302(EHS) TPQ	Section 304 EHS RQ	Section 304 CERCLA RQ	Sec 313	RCRA Code
16923-95-8	Zirconium potassium fluoride			1,000		
17702-41-9	Decaborane(14)	500/10,000	1			
17702-57-7	Formparanate	100/10,000	1			
18883-66-4	D-Glucose, 2-deoxy-2-[[(methylnitrosoamino			1		U206
19287-45-7	Diborane	100	1			
19624-22-7	Pentaborane	500	1			
20816-12-0	Osmium tetroxide			1,000	313	P087
20816-12-0	Osmium oxide OsO4 (T-4)-			1,000	X	P087
20830-75-5	Digoxin	10/10,000	1			
20830-81-3	Daunomycin			10		U059
20859-73-8	Aluminum phosphide	500		100		P006
21548-32-3	Fosthietan	500	1			
21609-90-5	Leptophos	500/10,000	1			
21908-53-2	Mercuric oxide	500/10,000	1			
21923-23-9	Chlorthiophos	500	1			
22224-92-6	Fenamiphos	10/10,000	1			
23135-22-0	Oxamyl	100/10,000	1			
23422-53-9	Formetanate hydrochloride	500/10,000	1			
23505-41-1	Pirimifos-ethyl	1,000	1			
23950-58-5	Benzamide,3,5-dichloro-N-(1,1-dimethyl-2-p			5,000		U192
24017-47-8	Triazofos	500	1			
24934-91-6	Chlormephos	500	1			
25154-54-5	Dinitrobenzene (mixed isomers)			100		
25154-55-6	Nitrophenol (mixed isomers)			100		
25155-30-0	Sodium dodecylbenzenesulfonate			1,000		
25167-82-2	Trichlorophenol			10		
25168-15-4	2,4,5-T esters			1,000		
25168-26-7	2,4-D Esters			100		
25321-14-6	Dinitrotoluene (mixed isomers)			10	313	
25321-22-6	Dichlorobenzene (mixed isomers)			100	313	
25321-22-6	Dichlorobenzene			100	X	
25376-45-8	Diaminotoluene (mixed isomers)			10	313	U221
25376-45-8	Toluenediamine			10	X	
25550-58-7	Dinitrophenol			10		
26264-06-2	Calcium dodecylbenzenesulfonate			1,000		
26419-73-8	Carbamic acid, methyl-, O-(((2,4-dimethyl-	100/10,000	1			
26471-62-5	Toluenediisocyanate (mixed isomers)			100	313	U223
26628-22-8	Sodium azide (Na(N3))	500		1,000		P105
26638-19-7	Dichloropropane			1,000		
27137-85-5	Trichloro(dichlorophenyl)silane	500	1			

CAS Number	Chemical Name	Sec. 302(EHS) TPQ	Section 304 EHS RQ	Section 304 CERCLA RQ	Sec 313	RCRA Code
27176-87-0	Dodecylbenzenesulfonic acid			1,000		
27323-41-7	Triethanolamine dodecylbenzene sulfonate			1,000		
27774-13-6	Vanadyl sulfate			1,000		
28300-74-5	Antimony potassium tartrate			100		
28347-13-9	Xylylene dichloride	100/10,000	1			
28772-56-7	Bromadiolone	100/10,000	1			
30525-89-4	Paraformaldehyde			1,000		
30674-80-7	Methacryloyloxyethyl isocyanate	100	1			
32534-95-5	2,4,5-TP esters			100		
33213-65-9	beta - Endosulfan			1		
36478-76-9	Uranyl nitrate			100		
37211-05-5	Nickel chloride			100		
39156-41-7	2,4-Diaminoanisole sulfate				313	
39196-18-4	Thiofanox	100/10,000		100		P045
42504-46-1	Isopropanolamine dodecylbenzene sulfonate			1,000		
50782-69-9	Phosphonothioic acid, methyl-, S-(2-(bis(1	100	1			
52628-25-8	Zinc ammonium chloride			1,000		
52652-59-2	Lead stearate			5,000		
52740-16-6	Calcium arsenite			1		
53467-11-1	2,4-D Esters			100		
53469-21-9	Aroclor 1242			1		
53558-25-1	Pyriminil	100/10,000	1			
55488-87-4	Ferric ammonium oxalate			1,000		
56189-09-4	Lead stearate			5,000		
58270-08-9	Zinc, dichloro(4,4-dimethyl-5(((methylami	100/10,000	1			
61792-07-2	2,4,5-T esters			1,000		
62207-76-5	Cobalt, ((2,2'-(1,2-ethanediylbis(nitrilom	100/10,000	1			
**	Organorhodium Complex (PMN-82-147)	10/10,000	1			

** This chemical was identified from a Premanufacture Review Notice (PMN) submitted to EPA. The submitter has claimed certain information on the submission to be confidential, including specific chemical identity.

Chemical Category	Sec. 302(EHS) TPQ	Section 304 EHS RQ	Section 304 CERCLA RQ	Sec 313	RCRA Code
Antimony Compounds			***	313	
Arsenic Compounds			***	313	
Barium Compounds				313	
Beryllium Compounds			***	313	
Cadmium Compounds			***	313	
Chlordane (Technical Mixture and Metabolites)			***		
Chlorinated Benzenes			***		
Chlorinated Ethanes			***		
Chlorinated Naphthalene			***		
Chlorophenols			***	313	
Chlorinated Phenols			***	X	
Chloroalkyl Ethers			***		
Chromium Compounds			***	313	
Cobalt Compounds			1+	313	
Coke Oven Emissions			1		
Copper Compounds			***	313	
Cyanide Compounds			***	313	
DDT and Metabolites			***		
Dichlorobenzidine			***		
Diphenylhydrazine			***		
Endosulfan and Metabolites			***		
Endrin and Metabolites			***		
Glycol Ethers			1+	313	
Haloethers			***		
Halomethanes			***		
Heptachlor and Metabolites			***		
Lead Compounds			***	313	
Manganese Compounds			1+	313	
Mercury Compounds			***	313	
Fine mineral fibers			1+		
Nickel Compounds			***	313	
Nitrophenols			***		
Nitrosamines			***		
Phthalate Esters			***		
Polybrominated Biphenyls (PBBs)				313	
Polycyclic organic matter			1+		
Polynuclear Aromatic Hydrocarbons			***		
Selenium Compounds			***	313	
Silver Compounds			***	313	
Thallium Compounds			***	313	
Zinc Compounds			***	313	

*** Indicates that no RQ is assigned to this generic or broad class, although the class is a CERCLA hazardous substance. See 50 Federal Register 13456 (April 4, 1985).

+ Listed as hazardous air pollutant under section 112(b) of the Clear Air Act; statutory RQ of one pound applies until RQs are adjusted.

ALPHABETICAL LISTING OF CHEMICAL NAME AND CAS NUMBER

CAS Number	Chemical Name
83-32-9	Acenaphthene
208-96-8	Acenaphthylene
75-07-0	Acetaldehyde
75-87-6	Acetaldehyde, trichloro-
60-35-5	Acetamide
64-19-7	Acetic acid
108-24-7	Acetic anhydride
67-64-1	Acetone
75-86-5	Acetone cyanohydrin
1752-30-3	Acetone thiosemicarbazide
75-05-8	Acetonitrile
98-86-2	Acetophenone
53-96-3	2-Acetylaminofluorene
506-96-7	Acetyl bromide
75-36-5	Acetyl chloride
591-08-2	1-Acetyl-2-thiourea
107-02-8	Acrolein
79-06-1	Acrylamide
79-10-7	Acrylic acid
107-13-1	Acrylonitrile
814-68-6	Acrylyl chloride
124-04-9	Adipic acid
111-69-3	Adiponitrile
116-06-3	Aldicarb
309-00-2	Aldrin
107-18-6	Allyl alcohol
107-11-9	Allylamine
107-05-1	Allyl chloride
7429-90-5	Aluminum (fume or dust)
1344-28-1	Aluminum oxide (fibrous forms)
20859-73-8	Aluminum phosphide
10043-01-3	Aluminum sulfate
117-79-3	2-Aminoanthraquinone
60-09-3	4-Aminoazobenzene
92-67-1	4-Aminobiphenyl
82-28-0	1-Amino-2-methylanthraquinone
54-62-6	Aminopterin
504-24-5	4-Aminopyridine
78-53-5	Amiton
3734-97-2	Amiton oxalate
61-82-5	Amitrole
7664-41-7	Ammonia
631-61-8	Ammonium acetate
1863-63-4	Ammonium benzoate
1066-33-7	Ammonium bicarbonate
7789-09-5	Ammonium bichromate
1341-49-7	Ammonium bifluoride
10192-30-0	Ammonium bisulfite
1111-78-0	Ammonium carbamate
506-87-6	Ammonium carbonate
12125-02-9	Ammonium chloride
7788-98-9	Ammonium chromate
3012-65-5	Ammonium citrate, dibasic
13826-83-0	Ammonium fluoborate
12125-01-8	Ammonium fluoride
1336-21-6	Ammonium hydroxide
6484-52-2	Ammonium nitrate (solution)
6009-70-7	Ammonium oxalate
5972-73-6	Ammonium oxalate
14258-49-2	Ammonium oxalate
131-74-8	Ammonium picrate
16919-19-0	Ammonium silicofluoride
7773-06-0	Ammonium sulfamate
7783-20-2	Ammonium sulfate (solution)
12135-76-1	Ammonium sulfide
10196-04-0	Ammonium sulfite
14307-43-8	Ammonium tartrate
3164-29-2	Ammonium tartrate
1762-95-4	Ammonium thiocyanate
7803-55-6	Ammonium vanadate
300-62-9	Amphetamine
628-63-7	Amyl acetate
123-92-2	iso-Amyl acetate
626-38-0	sec-Amyl acetate
625-16-1	tert-Amyl acetate
62-53-3	Aniline
88-05-1	Aniline, 2,4,6-trimethyl-
90-04-0	o-Anisidine
104-94-9	p-Anisidine
134-29-2	o-Anisidine hydrochloride
120-12-7	Anthracene
7440-36-0	Antimony
	Antimony Compounds
7647-18-9	Antimony pentachloride
7783-70-2	Antimony pentafluoride
28300-74-5	Antimony potassium tartrate
7789-61-9	Antimony tribromide
10025-91-9	Antimony trichloride
7783-56-4	Antimony trifluoride
1309-64-4	Antimony trioxide
1397-94-0	Antimycin A
86-88-4	Antu
12674-11-2	Aroclor 1016
11104-28-2	Aroclor 1221
11141-16-5	Aroclor 1232
53469-21-9	Aroclor 1242
12672-29-6	Aroclor 1248
11097-69-1	Aroclor 1254
11096-82-5	Aroclor 1260
7440-38-2	Arsenic

CAS Number	Chemical Name
1327-52-2	Arsenic acid
7778-39-4	Arsenic acid
	Arsenic Compounds
1303-32-8	Arsenic disulfide
1303-28-2	Arsenic pentoxide
1327-53-3	Arsenic trioxide
1303-33-9	Arsenic trisulfide
1327-53-3	Arsenous oxide
7784-34-1	Arsenous trichloride
7784-42-1	Arsine
1332-21-4	Asbestos (friable)
492-80-8	Auramine
115-02-6	Azaserine
2642-71-9	Azinphos-ethyl
86-50-0	Azinphos-methyl
151-56-4	Aziridine
75-55-8	Aziridine, 2-methyl
7440-39-3	Barium
	Barium Compounds
542-62-1	Barium cyanide
225-51-4	Benz[c]acridine
98-87-3	Benzal chloride
55-21-0	Benzamide
23950-58-5	Benzamide,3,5-dichloro-N-(1,1-dimethyl-2-p
56-55-3	Benz[a]anthracene
98-16-8	Benzenamine, 3-(trifluoromethyl)-
71-43-2	Benzene
98-05-5	Benzenearsonic acid
100-14-1	Benzene, 1-(chloromethyl)-4-nitro-
108-38-3	Benzene, m-dimethyl-
95-47-6	Benzene, o-dimethyl-
106-42-3	Benzene, p-dimethyl-
122-09-8	Benzeneethanamine, alpha,alpha-dimethyl-
98-09-9	Benzenesulfonyl chloride
108-98-5	Benzenethiol
92-87-5	Benzidine
3615-21-2	Benzimidazole, 4,5-dichloro-2-(trifluorome
205-99-2	Benzo[b]fluoranthene
207-08-9	Benzo(k)fluoranthene
65-85-0	Benzoic acid
98-07-7	Benzoic trichloride
100-47-0	Benzonitrile
191-24-2	Benzo[ghi]perylene
50-32-8	Benzo[a]pyrene
106-51-4	p-Benzoquinone
98-07-7	Benzotrichloride
98-88-4	Benzoyl chloride
94-36-0	Benzoyl peroxide
100-44-7	Benzyl chloride
140-29-4	Benzyl cyanide
7440-41-7	Beryllium
7787-47-5	Beryllium chloride
	Beryllium Compounds
7787-49-7	Beryllium fluoride
13597-99-4	Beryllium nitrate
7787-55-5	Beryllium nitrate
319-84-6	alpha-BHC
319-85-7	beta-BHC
319-86-8	delta-BHC
15271-41-7	Bicyclo[2.2.1]heptane-2-carbonitrile, 5-ch
1464-53-5	2,2'-Bioxirane
92-52-4	Biphenyl
111-91-1	Bis(2-chloroethoxy) methane
111-44-4	Bis(2-chloroethyl) ether
542-88-1	Bis(chloromethyl) ether
108-60-1	Bis(2-chloro-1-methylethyl)ether
534-07-6	Bis(chloromethyl) ketone
103-23-1	Bis(2-ethylhexyl) adipate
117-81-7	Bis(2-ethylhexyl)phthalate
4044-65-9	Bitoscanate
10294-34-5	Boron trichloride
7637-07-2	Boron trifluoride
353-42-4	Boron trifluoride compound with methyl eth
28772-56-7	Bromadiolone
7726-95-6	Bromine
598-31-2	Bromoacetone
353-59-3	Bromochlorodifluoromethane [Halon 1211]
75-25-2	Bromoform
74-83-9	Bromomethane
101-55-3	4-Bromophenyl phenyl ether
75-63-8	Bromotrifluoromethane [Halon 1301]
357-57-3	Brucine
106-99-0	1,3-Butadiene
764-41-0	2-Butene, 1,4-dichloro-
123-86-4	Butyl acetate
110-19-0	iso-Butyl acetate
105-46-4	sec-Butyl acetate
540-88-5	tert-Butyl acetate
141-32-2	Butyl acrylate
71-36-3	n-Butyl alcohol
78-92-2	sec-Butyl alcohol
75-65-0	tert-Butyl alcohol
109-73-9	Butylamine
78-81-9	iso-Butylamine
513-49-5	sec-Butylamine
13952-84-6	sec-Butylamine
75-64-9	tert-Butylamine
85-68-7	Butyl benzyl phthalate
106-88-7	1,2-Butylene oxide
84-74-2	n-Butyl phthalate

CAS Number	Chemical Name
123-72-8	Butyraldehyde
107-92-6	Butyric acid
79-31-2	iso-Butyric acid
75-60-5	Cacodylic acid
7440-43-9	Cadmium
543-90-8	Cadmium acetate
7789-42-6	Cadmium bromide
10108-64-2	Cadmium chloride
	Cadmium Compounds
1306-19-0	Cadmium oxide
2223-93-0	Cadmium stearate
7778-44-1	Calcium arsenate
52740-16-6	Calcium arsenite
75-20-7	Calcium carbide
13765-19-0	Calcium chromate
156-62-7	Calcium cyanamide
592-01-8	Calcium cyanide
26264-06-2	Calcium dodecylbenzenesulfonate
7778-54-3	Calcium hypochlorite
8001-35-2	Camphechlor
8001-35-2	Camphene, octachloro-
56-25-7	Cantharidin
105-60-2	Caprolactam
133-06-2	Captan
51-83-2	Carbachol chloride
51-79-6	Carbamic acid, ethyl ester
26419-73-8	Carbamic acid, methyl-, O-(((2,4-dimethyl-
63-25-2	Carbaryl
1563-66-2	Carbofuran
75-15-0	Carbon disulfide
353-50-4	Carbonic difluoride
56-23-5	Carbon tetrachloride
463-58-1	Carbonyl sulfide
786-19-6	Carbophenothion
120-80-9	Catechol
75-69-4	CFC-11
75-71-8	CFC-12
76-14-2	CFC-114
76-15-3	CFC-115
133-90-4	Chloramben
305-03-3	Chlorambucil
57-74-9	Chlordane
	Chlordane (Technical Mixture and Metabolit
470-90-6	Chlorfenvinfos
	Chlorinated Benzenes
	Chlorinated Ethanes
	Chlorinated Naphthalene
	Chlorinated Phenols
7782-50-5	Chlorine
10049-04-4	Chlorine dioxide
24934-91-6	Chlormephos
999-81-5	Chlormequat chloride
494-03-1	Chlornaphazine
107-20-0	Chloroacetaldehyde
79-11-8	Chloroacetic acid
532-27-4	2-Chloroacetophenone
	Chloroalkyl Ethers
106-47-8	p-Chloroaniline
108-90-7	Chlorobenzene
510-15-6	Chlorobenzilate
59-50-7	p-Chloro-m-cresol
124-48-1	Chlorodibromomethane
75-00-3	Chloroethane
107-07-3	Chloroethanol
627-11-2	Chloroethyl chloroformate
110-75-8	2-Chloroethyl vinyl ether
67-66-3	Chloroform
74-87-3	Chloromethane
107-30-2	Chloromethyl methyl ether
542-88-1	Chloromethyl ether
91-58-7	2-Chloronaphthalene
3691-35-8	Chlorophacinone
95-57-8	2-Chlorophenol
	Chlorophenols
7005-72-3	4-Chlorophenyl phenyl ether
126-99-8	Chloroprene
542-76-7	3-Chloropropionitrile
7790-94-5	Chlorosulfonic acid
1897-45-6	Chlorothalonil
3165-93-3	4-Chloro-o-toluidine, hydrochloride
1982-47-4	Chloroxuron
2921-88-2	Chlorpyrifos
21923-23-9	Chlorthiophos
1066-30-4	Chromic acetate
11115-74-5	Chromic acid
7738-94-5	Chromic acid
10025-73-7	Chromic chloride
10101-53-8	Chromic sulfate
7440-47-3	Chromium
	Chromium Compounds
10049-05-5	Chromous chloride
218-01-9	Chrysene
4680-78-8	C.I. Acid Green 3
569-64-2	C.I. Basic Green 4
989-38-8	C.I. Basic Red 1
1937-37-7	C.I. Direct Black 38
2602-46-2	C.I. Direct Blue 6
16071-86-6	C.I. Direct Brown 95
2832-40-8	C.I. Disperse Yellow 3
3761-53-3	C.I. Food Red 5

CAS Number	Chemical Name
81-88-9	C.I. Food Red 15
3118-97-6	C.I. Solvent Orange 7
97-56-3	C.I. Solvent Yellow 3
842-07-9	C.I. Solvent Yellow 14
492-80-8	C.I. Solvent Yellow 34
128-66-5	C.I. Vat Yellow 4
7440-48-4	Cobalt
10210-68-1	Cobalt carbonyl
	Cobalt Compounds
62207-76-5	Cobalt, ((2,2'-(1,2-ethanediylbis(nitrilom
7789-43-7	Cobaltous bromide
544-18-3	Cobaltous formate
14017-41-5	Cobaltous sulfamate
	Coke Oven Emissions
64-86-8	Colchicine
7440-50-8	Copper
	Copper Compounds
544-92-3	Copper cyanide
56-72-4	Coumaphos
5836-29-3	Coumatetralyl
8001-58-9	Creosote
120-71-8	p-Cresidine
108-39-4	m-Cresol
95-48-7	o-Cresol
106-44-5	p-Cresol
1319-77-3	Cresol (mixed isomers)
535-89-7	Crimidine
4170-30-3	Crotonaldehyde
123-73-9	Crotonaldehyde, (E)-
98-82-8	Cumene
80-15-9	Cumene hydroperoxide
135-20-6	Cupferron
142-71-2	Cupric acetate
12002-03-8	Cupric acetoarsenite
7447-39-4	Cupric chloride
3251-23-8	Cupric nitrate
5893-66-3	Cupric oxalate
7758-98-7	Cupric sulfate
10380-29-7	Cupric sulfate, ammoniated
815-82-7	Cupric tartrate
	Cyanide Compounds
57-12-5	Cyanides (soluble salts and complexes)
460-19-5	Cyanogen
506-68-3	Cyanogen bromide
506-77-4	Cyanogen chloride
506-78-5	Cyanogen iodide
2636-26-2	Cyanophos
675-14-9	Cyanuric fluoride
110-82-7	Cyclohexane
108-94-1	Cyclohexanone
66-81-9	Cycloheximide
108-91-8	Cyclohexylamine
131-89-5	2-Cyclohexyl-4,6-Dinitrophenol
50-18-0	Cyclophosphamide
94-75-7	2,4-D
94-75-7	2,4-D Acid
94-11-1	2,4-D Esters
94-79-1	2,4-D Esters
94-80-4	2,4-D Esters
1320-18-9	2,4-D Esters
1928-38-7	2,4-D Esters
1928-61-6	2,4-D Esters
1929-73-3	2,4-D Esters
2971-38-2	2,4-D Esters
25168-26-7	2,4-D Esters
53467-11-1	2,4-D Esters
94-75-7	2,4-D, salts and esters
20830-81-3	Daunomycin
96-12-8	DBCP
72-54-8	DDD
72-55-9	DDE
3547-04-4	DDE
50-29-3	DDT
	DDT and Metabolites
17702-41-9	Decaborane(14)
1163-19-5	Decabromodiphenyl oxide
117-81-7	DEHP
8065-48-3	Demeton
919-86-8	Demeton-S-methyl
10311-84-9	Dialifor
2303-16-4	Diallate
615-05-4	2,4-Diaminoanisole
39156-41-7	2,4-Diaminoanisole sulfate
101-80-4	4,4'-Diaminodiphenyl ether
496-72-0	Diaminotoluene
823-40-5	Diaminotoluene
95-80-7	2,4-Diaminotoluene
25376-45-8	Diaminotoluene (mixed isomers)
333-41-5	Diazinon
334-88-3	Diazomethane
53-70-3	Dibenz[a,h]anthracene
132-64-9	Dibenzofuran
189-55-9	Dibenz[a,i]pyrene
19287-45-7	Diborane
96-12-8	1,2-Dibromo-3-chloropropane
106-93-4	1,2-Dibromoethane
124-73-2	Dibromotetrafluoroethane [Halon 2402]
84-74-2	Dibutyl phthalate
1918-00-9	Dicamba
1194-65-6	Dichlobenil

CAS Number	Chemical Name
117-80-6	Dichlone
25321-22-6	Dichlorobenzene
95-50-1	o-Dichlorobenzene
95-50-1	1,2-Dichlorobenzene
541-73-1	1,3-Dichlorobenzene
106-46-7	1,4-Dichlorobenzene
25321-22-6	Dichlorobenzene (mixed isomers)
91-94-1	3,3'-Dichlorobenzidine
	Dichlorobenzidine
75-27-4	Dichlorobromomethane
110-57-6	Trans-1,4-dichlorobutene
75-71-8	Dichlorodifluoromethane [CFC-12]
107-06-2	1,2-Dichloroethane
75-34-3	1,1-Dichloroethane
540-59-0	1,2-Dichloroethylene
75-35-4	1,1-Dichloroethylene
156-60-5	1,2-Dichloroethylene
111-44-4	Dichloroethyl ether
108-60-1	Dichloroisopropyl ether
75-09-2	Dichloromethane
542-88-1	Dichloromethyl ether
149-74-6	Dichloromethylphenylsilane
120-83-2	2,4-Dichlorophenol
87-65-0	2,6-Dichlorophenol
696-28-6	Dichlorophenylarsine
78-87-5	1,2-Dichloropropane
26638-19-7	Dichloropropane
8003-19-8	Dichloropropane - Dichloropropene (mixture
78-99-9	1,1-Dichloropropane
142-28-9	1,3-Dichloropropane
542-75-6	1,3-Dichloropropene
78-88-6	2,3-Dichloropropene
75-99-0	2,2-Dichloropropionic acid
542-75-6	1,3-Dichloropropylene
76-14-2	Dichlorotetrafluoroethane [CFC-114]
62-73-7	Dichlorvos
115-32-2	Dicofol
141-66-2	Dicrotophos
60-57-1	Dieldrin
1464-53-5	Diepoxybutane
111-42-2	Diethanolamine
109-89-7	Diethylamine
692-42-2	Diethylarsine
1642-54-2	Diethylcarbamazine citrate
814-49-3	Diethyl chlorophosphate
117-81-7	Di(2-ethylhexyl) phthalate
311-45-5	Diethyl-p-nitrophenyl phosphate
84-66-2	Diethyl phthalate
297-97-2	O,O-Diethyl O-pyrazinyl phosphorothioate
56-53-1	Diethylstilbestrol
64-67-5	Diethyl sulfate
71-63-6	Digitoxin
2238-07-5	Diglycidyl ether
20830-75-5	Digoxin
94-58-6	Dihydrosafrole
55-91-4	Diisopropylfluorophosphate
115-26-4	Dimefox
60-51-5	Dimethoate
119-90-4	3,3'-Dimethoxybenzidine
124-40-3	Dimethylamine
60-11-7	4-Dimethylaminoazobenzene
60-11-7	Dimethylaminoazobenzene
121-69-7	N,N-Dimethylaniline
57-97-6	7,12-Dimethylbenz[a]anthracene
119-93-7	3,3'-Dimethylbenzidine
79-44-7	Dimethylcarbamyl chloride
75-78-5	Dimethyldichlorosilane
68-12-2	Dimethylformamide
57-14-7	1,1-Dimethyl hydrazine
57-14-7	Dimethylhydrazine
105-67-9	2,4-Dimethylphenol
99-98-9	Dimethyl-p-phenylenediamine
2524-03-0	Dimethyl phosphorochloridothioate
131-11-3	Dimethyl phthalate
77-78-1	Dimethyl sulfate
644-64-4	Dimetilan
25154-54-5	Dinitrobenzene (mixed isomers)
99-65-0	m-Dinitrobenzene
528-29-0	o-Dinitrobenzene
100-25-4	p-Dinitrobenzene
534-52-1	Dinitrocresol
534-52-1	4,6-Dinitro-o-cresol
534-52-1	4,6-Dinitro-o-cresol and salts
25550-58-7	Dinitrophenol
51-28-5	2,4-Dinitrophenol
329-71-5	2,5-Dinitrophenol
573-56-8	2,6-Dinitrophenol
25321-14-6	Dinitrotoluene (mixed isomers)
121-14-2	2,4-Dinitrotoluene
606-20-2	2,6-Dinitrotoluene
610-39-9	3,4-Dinitrotoluene
88-85-7	Dinoseb
1420-07-1	Dinoterb
117-84-0	n-Dioctylphthalate
117-84-0	Di-n-octyl phthalate
123-91-1	1,4-Dioxane
78-34-2	Dioxathion
82-66-6	Diphacinone
122-66-7	1,2-Diphenylhydrazine
	Diphenylhydrazine

CAS Number	Chemical Name
152-16-9	Diphosphoramide, octamethyl-
142-84-7	Dipropylamine
85-00-7	Diquat
2764-72-9	Diquat
298-04-4	Disulfoton
514-73-8	Dithiazanine iodide
541-53-7	Dithiobiuret
3288-58-2	O,O-Diethyl S-methyl dithiophosphate
330-54-1	Diuron
27176-87-0	Dodecylbenzenesulfonic acid
316-42-7	Emetine, dihydrochloride
115-29-7	Endosulfan
959-98-8	alpha - Endosulfan
33213-65-9	beta - Endosulfan
	Endosulfan and Metabolites
1031-07-8	Endosulfan sulfate
145-73-3	Endothall
2778-04-3	Endothion
72-20-8	Endrin
7421-93-4	Endrin aldehyde
	Endrin and Metabolites
106-89-8	Epichlorohydrin
51-43-4	Epinephrine
2104-64-5	EPN
50-14-6	Ergocalciferol
379-79-3	Ergotamine tartrate
1622-32-8	Ethanesulfonyl chloride, 2-chloro-
630-20-6	Ethane, 1,1,1,2-tetrachloro-
16752-77-5	Ethanimidothioic acid, N-[[methylamino)car
10140-87-1	Ethanol, 1,2-dichloro-, acetate
110-80-5	Ethanol, 2-ethoxy-
563-12-2	Ethion
13194-48-4	Ethoprophos
110-80-5	2-Ethoxyethanol
141-78-6	Ethyl acetate
140-88-5	Ethyl acrylate
100-41-4	Ethylbenzene
538-07-8	Ethylbis(2-chloroethyl)amine
51-79-6	Ethyl carbamate
75-00-3	Ethyl chloride
541-41-3	Ethyl chloroformate
107-12-0	Ethyl cyanide
74-85-1	Ethylene
111-54-6	Ethylenebisdithiocarbamic acid, salts & es
107-15-3	Ethylenediamine
60-00-4	Ethylenediamine-tetraacetic acid (EDTA)
106-93-4	Ethylene dibromide
107-06-2	Ethylene dichloride
371-62-0	Ethylene fluorohydrin
107-21-1	Ethylene glycol
151-56-4	Ethyleneimine
75-21-8	Ethylene oxide
96-45-7	Ethylene thiourea
60-29-7	Ethyl ether
97-63-2	Ethyl methacrylate
62-50-0	Ethyl methanesulfonate
542-90-5	Ethylthiocyanate
52-85-7	Famphur
22224-92-6	Fenamiphos
122-14-5	Fenitrothion
115-90-2	Fensulfothion
1185-57-5	Ferric ammonium citrate
2944-67-4	Ferric ammonium oxalate
55488-87-4	Ferric ammonium oxalate
7705-08-0	Ferric chloride
7783-50-8	Ferric fluoride
10421-48-4	Ferric nitrate
10028-22-5	Ferric sulfate
10045-89-3	Ferrous ammonium sulfate
7758-94-3	Ferrous chloride
7720-78-7	Ferrous sulfate
7782-63-0	Ferrous sulfate
4301-50-2	Fluenetil
2164-17-2	Fluometuron
206-44-0	Fluoranthene
86-73-7	Fluorene
7782-41-4	Fluorine
640-19-7	Fluoroacetamide
144-49-0	Fluoroacetic acid
62-74-8	Fluoroacetic acid, sodium salt
359-06-8	Fluoroacetyl chloride
51-21-8	Fluorouracil
944-22-9	Fonofos
50-00-0	Formaldehyde
107-16-4	Formaldehyde cyanohydrin
23422-53-9	Formetanate hydrochloride
64-18-6	Formic acid
2540-82-1	Formothion
17702-57-7	Formparanate
21548-32-3	Fosthietan
76-13-1	Freon 113
3878-19-1	Fuberidazole
110-17-8	Fumaric acid
110-00-9	Furan
109-99-9	Furan, tetrahydro-
98-01-1	Furfural
13450-90-3	Gallium trichloride
18883-66-4	D-Glucose, 2-deoxy-2-[[(methylnitrosoamino
765-34-4	Glycidylaldehyde
	Glycol Ethers

CAS Number	Chemical Name
70-25-7	Guanidine, N-methyl-N'-nitro-N-nitroso-
86-50-0	Guthion
	Haloethers
	Halomethanes
353-59-3	Halon 1211
75-63-8	Halon 1301
124-73-2	Halon 2402
76-44-8	Heptachlor
	Heptachlor and Metabolites
1024-57-3	Heptachlor epoxide
118-74-1	Hexachlorobenzene
87-68-3	Hexachloro-1,3-butadiene
87-68-3	Hexachlorobutadiene
77-47-4	Hexachlorocyclopentadiene
58-89-9	Hexachlorocyclohexane (gamma isomer)
67-72-1	Hexachloroethane
1335-87-1	Hexachloronaphthalene
70-30-4	Hexachlorophene
1888-71-7	Hexachloropropene
757-58-4	Hexaethyl tetraphosphate
822-06-0	Hexamethylene-1,6-diisocyanate
4835-11-4	Hexamethylenediamine, N,N'-dibutyl-
680-31-9	Hexamethylphosphoramide
110-54-3	Hexane
302-01-2	Hydrazine
1615-80-1	Hydrazine, 1,2-diethyl-
57-14-7	Hydrazine, 1,1-dimethyl-
540-73-8	Hydrazine, 1,2-dimethyl-
122-66-7	Hydrazine, 1,2-diphenyl-
10034-93-2	Hydrazine sulfate
122-66-7	Hydrazobenzene
7647-01-0	Hydrochloric acid
74-90-8	Hydrocyanic acid
7664-39-3	Hydrofluoric acid
7647-01-0	Hydrogen chloride (gas only)
74-90-8	Hydrogen cyanide
7664-39-3	Hydrogen fluoride
7722-84-1	Hydrogen peroxide (Conc.> 52%)
7783-07-5	Hydrogen selenide
7783-06-4	Hydrogen sulfide
80-15-9	Hydroperoxide, 1-methyl-1-phenylethyl-
123-31-9	Hydroquinone
193-39-5	Indeno(1,2,3-cd)pyrene
13463-40-6	Iron, pentacarbonyl-
297-78-9	Isobenzan
78-83-1	Isobutyl alcohol
78-84-2	Isobutyraldehyde
78-82-0	Isobutyronitrile
102-36-3	Isocyanic acid, 3,4-dichlorophenyl ester
465-73-6	Isodrin
55-91-4	Isofluorphate
78-59-1	Isophorone
4098-71-9	Isophorone diisocyanate
78-79-5	Isoprene
42504-46-1	Isopropanolamine dodecylbenzene sulfonate
67-63-0	Isopropyl alcohol (mfg-strong acid process
108-23-6	Isopropyl chloroformate
80-05-7	4,4'-Isopropylidenediphenol
119-38-0	Isopropylmethylpyrazolyl dimethylcarbamate
120-58-1	Isosafrole
2763-96-4	5-(Aminomethyl)-3-isoxazolol
143-50-0	Kepone
78-97-7	Lactonitrile
303-34-4	Lasiocarpine
7439-92-1	Lead
	Lead Compounds
301-04-2	Lead acetate
7784-40-9	Lead arsenate
7645-25-2	Lead arsenate
10102-48-4	Lead arsenate
7758-95-4	Lead chloride
13814-96-5	Lead fluoborate
7783-46-2	Lead fluoride
10101-63-0	Lead iodide
10099-74-8	Lead nitrate
7446-27-7	Lead phosphate
7428-48-0	Lead stearate
1072-35-1	Lead stearate
52652-59-2	Lead stearate
56189-09-4	Lead stearate
1335-32-6	Lead subacetate
15739-80-7	Lead sulfate
7446-14-2	Lead sulfate
1314-87-0	Lead sulfide
592-87-0	Lead thiocyanate
21609-90-5	Leptophos
541-25-3	Lewisite
58-89-9	Lindane
14307-35-8	Lithium chromate
7580-67-8	Lithium hydride
121-75-5	Malathion
110-16-7	Maleic acid
108-31-6	Maleic anhydride
123-33-1	Maleic hydrazide
109-77-3	Malononitrile
12427-38-2	Maneb
7439-96-5	Manganese
	Manganese Compounds
12108-13-3	Manganese, tricarbonyl methylcyclopentadie
101-68-8	MBI

CAS Number	Chemical Name
101-14-4	MBOCA
51-75-2	Mechlorethamine
148-82-3	Melphalan
950-10-7	Mephosfolan
2032-65-7	Mercaptodimethur
1600-27-7	Mercuric acetate
7487-94-7	Mercuric chloride
592-04-1	Mercuric cyanide
10045-94-0	Mercuric nitrate
21908-53-2	Mercuric oxide
7783-35-9	Mercuric sulfate
592-85-8	Mercuric thiocyanate
10415-75-5	Mercurous nitrate
7782-86-7	Mercurous nitrate
7439-97-6	Mercury
	Mercury Compounds
628-86-4	Mercury fulminate
10476-95-6	Methacrolein diacetate
760-93-0	Methacrylic anhydride
126-98-7	Methacrylonitrile
920-46-7	Methacryloyl chloride
30674-80-7	Methacryloyloxyethyl isocyanate
10265-92-6	Methamidophos
62-75-9	Methanamine, N-methyl-N-nitroso-
558-25-8	Methanesulfonyl fluoride
67-56-1	Methanol
91-80-5	Methapyrilene
950-37-8	Methidathion
2032-65-7	Methiocarb
16752-77-5	Methomyl
72-43-5	Methoxychlor
109-86-4	2-Methoxyethanol
151-38-2	Methoxyethylmercuric acetate
96-33-3	Methyl acrylate
74-83-9	Methyl bromide
74-87-3	Methyl chloride
80-63-7	Methyl 2-chloroacrylate
71-55-6	Methyl chloroform
79-22-1	Methyl chloroformate
56-49-5	3-Methylcholanthrene
101-14-4	4,4'-Methylenebis(2-chloroaniline)
101-61-1	4,4'-Methylenebis(N,N-dimethyl)benzenamine
101-68-8	Methylenebis(phenylisocyanate)
74-95-3	Methylene bromide
75-09-2	Methylene chloride
101-77-9	4,4'-Methylenedianiline
78-93-3	Methyl ethyl ketone
78-93-3	Methyl ethyl ketone (MEK)
1338-23-4	Methyl ethyl ketone peroxide
60-34-4	Methyl hydrazine
74-88-4	Methyl iodide
108-10-1	Methyl isobutyl ketone
624-83-9	Methyl isocyanate
556-61-6	Methyl isothiocyanate
74-93-1	Methyl mercaptan
502-39-6	Methylmercuric dicyanamide
80-62-6	Methyl methacrylate
298-00-0	Methyl parathion
3735-23-7	Methyl phenkapton
676-97-1	Methyl phosphonic dichloride
1634-04-4	Methyl tert-butyl ether
556-64-9	Methyl thiocyanate
56-04-2	Methylthiouracil
75-79-6	Methyltrichlorosilane
78-94-4	Methyl vinyl ketone
1129-41-5	Metolcarb
7786-34-7	Mevinphos
315-18-4	Mexacarbate
90-94-8	Michler's ketone
	Fine mineral fibers
50-07-7	Mitomycin C
1313-27-5	Molybdenum trioxide
76-15-3	Monochloropentafluoroethane [CFC-115]
6923-22-4	Monocrotophos
75-04-7	Monoethylamine
74-89-5	Monomethylamine
2763-96-4	Muscimol
505-60-2	Mustard gas
300-76-5	Naled
91-20-3	Naphthalene
1338-24-5	Naphthenic acid
130-15-4	1,4-Naphthoquinone
134-32-7	alpha-Naphthylamine
91-59-8	beta-Naphthylamine
7440-02-0	Nickel
	Nickel Compounds
15699-18-0	Nickel ammonium sulfate
13463-39-3	Nickel carbonyl
7718-54-9	Nickel chloride
37211-05-5	Nickel chloride
557-19-7	Nickel cyanide
12054-48-7	Nickel hydroxide
14216-75-2	Nickel nitrate
7786-81-4	Nickel sulfate
54-11-5	Nicotine
54-11-5	Nicotine and salts
65-30-5	Nicotine sulfate
7697-37-2	Nitric acid
10102-43-9	Nitric oxide
139-13-9	Nitrilotriacetic acid

CAS Number	Chemical Name
100-01-6	p-Nitroaniline
99-59-2	5-Nitro-o-anisidine
98-95-3	Nitrobenzene
92-93-3	4-Nitrobiphenyl
1122-60-7	Nitrocyclohexane
1836-75-5	Nitrofen
10102-44-0	Nitrogen dioxide
10544-72-6	Nitrogen dioxide
51-75-2	Nitrogen mustard
55-63-0	Nitroglycerin
25154-55-6	Nitrophenol (mixed isomers)
554-84-7	m-Nitrophenol
100-02-7	p-Nitrophenol
88-75-5	2-Nitrophenol
100-02-7	4-Nitrophenol
	Nitrophenols
79-46-9	2-Nitropropane
	Nitrosamines
924-16-3	N-Nitrosodi-n-butylamine
1116-54-7	N-Nitrosodiethanolamine
55-18-5	N-Nitrosodiethylamine
62-75-9	N-Nitrosodimethylamine
62-75-9	Nitrosodimethylamine
86-30-6	N-Nitrosodiphenylamine
156-10-5	p-Nitrosodiphenylamine
621-64-7	N-Nitrosodi-n-propylamine
759-73-9	N-Nitroso-N-ethylurea
684-93-5	N-Nitroso-N-methylurea
4549-40-0	N-Nitrosomethylvinylamine
59-89-2	N-Nitrosomorpholine
615-53-2	N-Nitroso-N-methylurethane
16543-55-8	N-Nitrosonornicotine
100-75-4	N-Nitrosopiperidine
930-55-2	N-Nitrosopyrrolidine
1321-12-6	Nitrotoluene
99-08-1	m-Nitrotoluene
88-72-2	o-Nitrotoluene
99-99-0	p-Nitrotoluene
99-55-8	5-Nitro-o-toluidine
991-42-4	Norbormide
2234-13-1	Octachloronaphthalene
	Organorhodium Complex (PMN-82-147)
20816-12-0	Osmium oxide OsO4 (T-4)-
20816-12-0	Osmium tetroxide
630-60-4	Ouabain
23135-22-0	Oxamyl
78-71-7	Oxetane, 3,3-bis(chloromethyl)-
75-21-8	Oxirane
2497-07-6	Oxydisulfoton
10028-15-6	Ozone

CAS Number	Chemical Name
30525-89-4	Paraformaldehyde
123-63-7	Paraldehyde
1910-42-5	Paraquat
2074-50-2	Paraquat methosulfate
56-38-2	Parathion
298-00-0	Parathion-methyl
12002-03-8	Paris green
1336-36-3	PCBs
82-68-8	PCNB
19624-22-7	Pentaborane
608-93-5	Pentachlorobenzene
76-01-7	Pentachloroethane
82-68-8	Pentachloronitrobenzene
87-86-5	PCP
87-86-5	Pentachlorophenol
2570-26-5	Pentadecylamine
504-60-9	1,3-Pentadiene
79-21-0	Peracetic acid
127-18-4	Perchloroethylene
594-42-3	Perchloromethylmercaptan
62-44-2	Phenacetin
85-01-8	Phenanthrene
108-95-2	Phenol
64-00-6	Phenol, 3-(1-methylethyl)-, methylcarbamat
4418-66-0	Phenol, 2,2'-thiobis[4-chloro-6-methyl-
58-36-6	Phenoxarsine, 10,10'-oxydi-
696-28-6	Phenyl dichloroarsine
106-50-3	p-Phenylenediamine
59-88-1	Phenylhydrazine hydrochloride
62-38-4	Phenylmercuric acetate
62-38-4	Phenylmercury acetate
90-43-7	2-Phenylphenol
2097-19-0	Phenylsilatrane
103-85-5	Phenylthiourea
298-02-2	Phorate
4104-14-7	Phosacetim
947-02-4	Phosfolan
75-44-5	Phosgene
732-11-6	Phosmet
13171-21-6	Phosphamidon
7803-51-2	Phosphine
2703-13-1	Phosphonothioic acid, methyl-, O-ethyl O-(
50782-69-9	Phosphonothioic acid, methyl-, S-(2-(bis(1
2665-30-7	Phosphonothioic acid, methyl-, O-(4-nitrop
7664-38-2	Phosphoric acid
3254-63-5	Phosphoric acid, dimethyl 4-(methylthio) p
2587-90-8	Phosphorothioic acid, O,O-dimethyl-5-(2-(m
7723-14-0	Phosphorus (yellow or white)
7723-14-0	Phosphorus
10025-87-3	Phosphorus oxychloride

CAS Number	Chemical Name
10026-13-8	Phosphorus pentachloride
1314-56-3	Phosphorus pentoxide
7719-12-2	Phosphorus trichloride
	Phthalate Esters
85-44-9	Phthalic anhydride
57-47-6	Physostigmine
57-64-7	Physostigmine, salicylate (1:1)
109-06-8	2-Picoline
88-89-1	Picric acid
124-87-8	Picrotoxin
110-89-4	Piperidine
23505-41-1	Pirimifos-ethyl
	Polybrominated Biphenyls (PBBs)
1336-36-3	Polychlorinated biphenyls
	Polycyclic organic matter
	Polynuclear Aromatic Hydrocarbons
7784-41-0	Potassium arsenate
10124-50-2	Potassium arsenite
7778-50-9	Potassium bichromate
7789-00-6	Potassium chromate
151-50-8	Potassium cyanide
1310-58-3	Potassium hydroxide
7722-64-7	Potassium permanganate
506-61-6	Potassium silver cyanide
2631-37-0	Promecarb
78-87-5	Propane 1,2-dichloro-
1120-71-4	1,3-Propane sultone
1120-71-4	Propane sultone
2312-35-8	Propargite
107-19-7	Propargyl alcohol
106-96-7	Propargyl bromide
57-57-8	beta-Propiolactone
123-38-6	Propionaldehyde
79-09-4	Propionic acid
123-62-6	Propionic anhydride
107-12-0	Propionitrile
542-76-7	Propionitrile, 3-chloro-
70-69-9	Propiophenone, 4'-amino
114-26-1	Propoxur
109-61-5	Propyl chloroformate
107-10-8	n-Propylamine
115-07-1	Propylene (Propene)
75-55-8	Propyleneimine
75-56-9	Propylene oxide
621-64-7	Di-n-propylnitrosamine
2275-18-5	Prothoate
129-00-0	Pyrene
121-29-9	Pyrethrins
121-21-1	Pyrethrins
8003-34-7	Pyrethrins
110-86-1	Pyridine
504-24-5	Pyridine, 4-amino-
54-11-5	Pyridine, 3-(1-methyl-2-pyrrolidinyl)-,(S)
140-76-1	Pyridine, 2-methyl-5-vinyl-
1124-33-0	Pyridine, 4-nitro-, 1-oxide
53558-25-1	Pyriminil
91-22-5	Quinoline
106-51-4	Quinone
82-68-8	Quintozene
50-55-5	Reserpine
108-46-3	Resorcinol
81-07-2	Saccharin (manufacturing)
81-07-2	Saccharin and salts
94-59-7	Safrole
14167-18-1	Salcomine
107-44-8	Sarin
7783-00-8	Selenious acid
12039-52-0	Selenious acid, dithallium(1+) salt
7782-49-2	Selenium
	Selenium Compounds
7446-08-4	Selenium dioxide
7791-23-3	Selenium oxychloride
7488-56-4	Selenium sulfide
630-10-4	Selenourea
563-41-7	Semicarbazide hydrochloride
3037-72-7	Silane, (4-aminobutyl)diethoxymethyl-
7440-22-4	Silver
	Silver Compounds
506-64-9	Silver cyanide
7761-88-8	Silver nitrate
93-72-1	Silvex (2,4,5-TP)
7440-23-5	Sodium
7631-89-2	Sodium arsenate
7784-46-5	Sodium arsenite
26628-22-8	Sodium azide (Na(N3))
10588-01-9	Sodium bichromate
1333-83-1	Sodium bifluoride
7631-90-5	Sodium bisulfite
124-65-2	Sodium cacodylate
7775-11-3	Sodium chromate
143-33-9	Sodium cyanide (Na(CN))
25155-30-0	Sodium dodecylbenzenesulfonate
7681-49-4	Sodium fluoride
62-74-8	Sodium fluoroacetate
16721-80-5	Sodium hydrosulfide
1310-73-2	Sodium hydroxide
7681-52-9	Sodium hypochlorite
10022-70-5	Sodium hypochlorite
124-41-4	Sodium methylate
7632-00-0	Sodium nitrite

CAS Number	Chemical Name
7558-79-4	Sodium phosphate, dibasic
10039-32-4	Sodium phosphate, dibasic
10140-65-5	Sodium phosphate, dibasic
7601-54-9	Sodium phosphate, tribasic
7758-29-4	Sodium phosphate, tribasic
7785-84-4	Sodium phosphate, tribasic
10101-89-0	Sodium phosphate, tribasic
10124-56-8	Sodium phosphate, tribasic
10361-89-4	Sodium phosphate, tribasic
13410-01-0	Sodium selenate
10102-18-8	Sodium selenite
7782-82-3	Sodium selenite
10102-20-2	Sodium tellurite
900-95-8	Stannane, acetoxytriphenyl-
7789-06-2	Strontium chromate
57-24-9	Strychnine
57-24-9	Strychnine, and salts
60-41-3	Strychnine, sulfate
100-42-5	Styrene
96-09-3	Styrene oxide
3689-24-5	Sulfotep
3569-57-1	Sulfoxide, 3-chloropropyl octyl
7446-09-5	Sulfur dioxide
7664-93-9	Sulfuric acid
8014-95-7	Sulfuric acid (fuming)
12771-08-3	Sulfur monochloride
1314-80-3	Sulfur phosphide
7783-60-0	Sulfur tetrafluoride
7446-11-9	Sulfur trioxide
93-76-5	2,4,5-T acid
2008-46-0	2,4,5-T amines
1319-72-8	2,4,5-T amines
3813-14-7	2,4,5-T amines
6369-96-6	2,4,5-T amines
6369-97-7	2,4,5-T amines
93-79-8	2,4,5-T esters
1928-47-8	2,4,5-T esters
2545-59-7	2,4,5-T esters
25168-15-4	2,4,5-T esters
61792-07-2	2,4,5-T esters
13560-99-1	2,4,5-T salts
77-81-6	Tabun
13494-80-9	Tellurium
7783-80-4	Tellurium hexafluoride
107-49-3	Tepp
13071-79-9	Terbufos
95-94-3	1,2,4,5-Tetrachlorobenzene
1746-01-6	2,3,7,8-Tetrachlorodibenzo-p-dioxin (TCDD)
79-34-5	1,1,2,2-Tetrachloroethane
127-18-4	Tetrachloroethylene
58-90-2	2,3,4,6-Tetrachlorophenol
961-11-5	Tetrachlorvinphos
3689-24-5	Tetraethyldithiopyrophosphate
78-00-2	Tetraethyl lead
107-49-3	Tetraethyl pyrophosphate
597-64-8	Tetraethyltin
75-74-1	Tetramethyllead
509-14-8	Tetranitromethane
1314-32-5	Thallic oxide
7440-28-0	Thallium
	Thallium Compounds
563-68-8	Thallium(I) acetate
6533-73-9	Thallium(I) carbonate
7791-12-0	Thallium chloride TlCl
10102-45-1	Thallium(I) nitrate
10031-59-1	Thallium sulfate
7446-18-6	Thallium(I) sulfate
6533-73-9	Thallous carbonate
7791-12-0	Thallous chloride
2757-18-8	Thallous malonate
7446-18-6	Thallous sulfate
62-55-5	Thioacetamide
2231-57-4	Thiocarbazide
139-65-1	4,4'-Thiodianiline
39196-18-4	Thiofanox
74-93-1	Thiomethanol
297-97-2	Thionazin
108-98-5	Thiophenol
79-19-6	Thiosemicarbazide
62-56-6	Thiourea
5344-82-1	Thiourea, (2-chlorophenyl)-
614-78-8	Thiourea, (2-methylphenyl)-
86-88-4	Thiourea, 1-naphthalenyl-
137-26-8	Thiram
1314-20-1	Thorium dioxide
7550-45-0	Titanium tetrachloride
119-93-7	o-Tolidine
108-88-3	Toluene
25376-45-8	Toluenediamine
91-08-7	Toluene-2,6-diisocyanate
584-84-9	Toluene-2,4-diisocyanate
26471-62-5	Toluenediisocyanate (mixed isomers)
95-53-4	o-Toluidine
106-49-0	p-Toluidine
636-21-5	o-Toluidine hydrochloride
8001-35-2	Toxaphene
32534-95-5	2,4,5-TP esters
1031-47-6	Triamiphos
68-76-8	Triaziquone
24017-47-8	Triazofos

CAS Number	Chemical Name
75-25-2	Tribromomethane
52-68-6	Trichlorfon
76-02-8	Trichloroacetyl chloride
120-82-1	1,2,4-Trichlorobenzene
1558-25-4	Trichloro(chloromethyl)silane
27137-85-5	Trichloro(dichlorophenyl)silane
71-55-6	1,1,1-Trichloroethane
79-00-5	1,1,2-Trichloroethane
79-01-6	Trichloroethylene
115-21-9	Trichloroethylsilane
75-69-4	Trichlorofluoromethane [CFC-11]
594-42-3	Trichloromethanesulfenyl chloride
75-69-4	Trichloromonofluoromethane
327-98-0	Trichloronate
25167-82-2	Trichlorophenol
15950-66-0	2,3,4-Trichlorophenol
933-78-8	2,3,5-Trichlorophenol
933-75-5	2,3,6-Trichlorophenol
95-95-4	2,4,5-Trichlorophenol
88-06-2	2,4,6-Trichlorophenol
609-19-8	3,4,5-Trichlorophenol
98-13-5	Trichlorophenylsilane
27323-41-7	Triethanolamine dodecylbenzene sulfonate
998-30-1	Triethoxysilane
121-44-8	Triethylamine
1582-09-8	Trifluralin
75-50-3	Trimethylamine
95-63-6	1,2,4-Trimethylbenzene
75-77-4	Trimethylchlorosilane
824-11-3	Trimethylolpropane phosphite
540-84-1	2,2,4-Trimethylpentane
1066-45-1	Trimethyltin chloride
99-35-4	1,3,5-Trinitrobenzene
639-58-7	Triphenyltin chloride
555-77-1	Tris(2-chloroethyl)amine
126-72-7	Tris(2,3-dibromopropyl) phosphate
72-57-1	Trypan blue
66-75-1	Uracil mustard
541-09-3	Uranyl acetate
10102-06-4	Uranyl nitrate
36478-76-9	Uranyl nitrate
51-79-6	Urethane
2001-95-8	Valinomycin
7440-62-2	Vanadium (fume or dust)
1314-62-1	Vanadium pentoxide
27774-13-6	Vanadyl sulfate
108-05-4	Vinyl acetate monomer
108-05-4	Vinyl acetate
593-60-2	Vinyl bromide
75-01-4	Vinyl chloride
75-35-4	Vinylidene chloride
81-81-2	Warfarin
81-81-2	Warfarin, & salts, conc.>0.3%
129-06-6	Warfarin sodium
108-38-3	m-Xylene
95-47-6	o-Xylene
106-42-3	p-Xylene
1330-20-7	Xylene (mixed isomers)
1300-71-6	Xylenol
87-62-7	2,6-Xylidine
28347-13-9	Xylylene dichloride
7440-66-6	Zinc (fume or dust)
7440-66-6	Zinc
	Zinc Compounds
557-34-6	Zinc acetate
52628-25-8	Zinc ammonium chloride
14639-97-5	Zinc ammonium chloride
14639-98-6	Zinc ammonium chloride
1332-07-6	Zinc borate
7699-45-8	Zinc bromide
3486-35-9	Zinc carbonate
7646-85-7	Zinc chloride
557-21-1	Zinc cyanide
58270-08-9	Zinc, dichloro(4,4-dimethyl-5(((methylami
7783-49-5	Zinc fluoride
557-41-5	Zinc formate
7779-86-4	Zinc hydrosulfite
7779-88-6	Zinc nitrate
127-82-2	Zinc phenolsulfonate
1314-84-7	Zinc phosphide
1314-84-7	Zinc phosphide (conc. <= 10%)
1314-84-7	Zinc phosphide (conc. > 10%)
16871-71-9	Zinc silicofluoride
7733-02-0	Zinc sulfate
12122-67-7	Zineb
13746-89-9	Zirconium nitrate
16923-95-8	Zirconium potassium fluoride
14644-61-2	Zirconium sulfate
10026-11-6	Zirconium tetrachloride

RADIONUCLIDES LISTED UNDER CERCLA
FOR REFERENCE ONLY, NOT FOR REGULATORY COMPLIANCE
SEE 40 CFR PART 302, TABLE 302.4, APPENDIX B, FOR MORE INFORMATION

Radionuclide Name	Atomic Number	RQ (curies)
Radionuclides (unlisted)		1
Actinium-224	89	100
Actinium-225	89	1
Actinium-225	89	1
Actinium-226	89	10
Actinium-227	89	0.001
Actinium-228	89	10
Aluminum-026	13	10
Americium-237	95	1000
Americium-238	95	100
Americium-239	95	100
Americium-240	95	10
Americium-241	95	0.01
Americium-242	95	100
Americium-242m	95	0.01
Americium-243	95	0.01
Americium-244	95	10
Americium-244m	95	1000
Americium-245	95	1000
Americium-246	95	1000
Americium-246m	95	1000
Antimony-115	51	1000
Antimony-116	51	1000
Antimony-116m	51	100
Antimony-117	51	1000
Antimony-118m	51	10
Antimony-119	51	1000
Antimony-120 (16 min)	51	1000
Antimony-120 (5.76 day)	51	10
Antimony-122	51	10
Antimony-124	51	10
Antimony-124m	51	1000
Antimony-125	51	10
Antimony-126	51	10
Antimony-126m	51	1000
Antimony-127	51	10
Antimony-128 (10.4 min)	51	1000
Antimony-128 (9.01 hours)	51	10
Antimony-129	51	100
Antimony-130	51	100
Antimony-131	51	1000
Argon-039	18	1000
Argon-041	18	10
Arsenic-069	33	1000
Arsenic-070	33	100
Arsenic-071	33	100
Arsenic-072	33	10
Arsenic-073	33	100
Arsenic-074	33	10
Arsenic-076	33	100
Arsenic-077	33	1000
Arsenic-078	33	100
Astatine-207	85	100
Astatine-211	85	100
Barium-126	56	1000
Barium-128	56	10
Barium-131	56	10
Barium-131m	56	1000
Barium-133	56	10
Barium-133m	56	100
Barium-135m	56	1000
Barium-139	56	1000
Barium-140	56	10
Barium-141	56	1000
Barium-142	56	1000
Berkelium-245	97	100
Berkelium-246	97	10
Berkelium-247	97	0.01
Berkelium-249	97	1
Berkelium-250	97	100
Beryllium-007	4	100
Beryllium-010	4	1
Bismuth-200	83	100
Bismuth-201	83	100
Bismuth-202	83	1000
Bismuth-203	83	10
Bismuth-205	83	10
Bismuth-206	83	10
Bismuth-207	83	10
Bismuth-210	83	10
Bismuth-210m	83	0.1
Bismuth-212	83	100
Bismuth-213	83	100
Bismuth-214	83	100
Bromine-074	35	100
Bromine-074m	35	100
Bromine-075	35	100
Bromine-076	35	10
Bromine-077	35	100
Bromine-080	35	1000
Bromine-080m	35	1000
Bromine-082	35	10
Bromine-083	35	1000
Bromine-084	35	100
Cadmium-104	48	1000
Cadmium-107	48	1000
Cadmium-109	48	1
Cadmium-113	48	0.1
Cadmium-113m	48	0.1
Cadmium-115	48	100

Radionuclide Name	Atomic Number	RQ (curies)	Radionuclide Name	Atomic Number	RQ (curies)
Cadmium-115m	48	10	Cobalt-060m	27	1000
Cadmium-117	48	100	Cobalt-061	27	1000
Cadmium-117m	48	10	Cobalt-062m	27	1000
Calcium-041	20	10	Copper-060	29	100
Calcium-045	20	10	Copper-061	29	100
Calcium-047	20	10	Copper-064	29	1000
Californium-244	98	1000	Copper-067	29	100
Californium-246	98	10	Curium-238	96	1000
Californium-248	98	0.1	Curium-240	96	1
Californium-249	98	0.01	Curium-241	96	10
Californium-250	98	0.01	Curium-242	96	1
Californium-251	98	0.01	Curium-243	96	0.01
Californium-252	98	0.1	Curium-244	96	0.01
Californium-253	98	10	Curium-245	96	0.01
Californium-254	98	0.1	Curium-246	96	0.01
Carbon-011	6	1000	Curium-247	96	0.01
Carbon-014	6	10	Curium-248	96	0.001
Cerium-134	58	10	Curium-249	96	1000
Cerium-135	58	10	Dysprosium-155	66	100
Cerium-137	58	1000	Dysprosium-157	66	100
Cerium-137m	58	100	Dysprosium-159	66	100
Cerium-139	58	100	Dysprosium-165	66	1000
Cerium-141	58	10	Dysprosium-166	66	10
Cerium-143	58	100	Einsteinium-250	99	10
Cerium-144	58	1	Einsteinium-251	99	1000
Cesium-125	55	1000	Einsteinium-253	99	10
Cesium-127	55	100	Einsteinium-254	99	0.1
Cesium-129	55	100	Einsteinium-254m	99	1
Cesium-130	55	1000	Erbium-161	68	100
Cesium-131	55	1000	Erbium-165	68	1000
Cesium-132	55	10	Erbium-169	68	100
Cesium-134	55	1	Erbium-171	68	100
Cesium-134m	55	1000	Erbium-172	68	10
Cesium-135	55	10	Europium-145	63	10
Cesium-135m	55	100	Europium-146	63	10
Cesium-136	55	10	Europium-147	63	10
Cesium-137	55	1	Europium-148	63	10
Cesium-138	55	100	Europium-149	63	100
Chlorine-036	17	10	Europium-150	63	1000
Chlorine-038	17	100	Europium-150	63	10
Chlorine-039	17	100	Europium-152	63	10
Chromium-048	24	100	Europium-152m	63	100
Chromium-049	24	1000	Europium-154	63	10
Chromium-051	24	1000	Europium-155	63	10
Cobalt-055	27	10	Europium-156	63	10
Cobalt-056	27	10	Europium-157	63	10
Cobalt-057	27	100	Europium-158	63	1000
Cobalt-058	27	10	Fermium-252	100	10
Cobalt-058m	27	1000	Fermium-253	100	10
Cobalt-060	27	10	Fermium-254	100	100

Radionuclide Name	Atomic Number	RQ (curies)
Fermium-255	100	100
Fermium-257	100	100
Fluorine-018	9	1000
Francium-222	87	100
Francium-223	87	100
Gadolinium-145	64	100
Gadolinium-146	64	10
Gadolinium-147	64	10
Gadolinium-148	64	0.001
Gadolinium-149	64	100
Gadolinium-151	64	100
Gadolinium-152	64	0.001
Gadolinium-153	64	10
Gadolinium-159	64	1000
Gallium-065	31	1000
Gallium-066	31	10
Gallium-067	31	100
Gallium-068	31	1000
Gallium-070	31	1000
Gallium-072	31	10
Gallium-073	31	100
Germanium-066	32	100
Germanium-067	32	1000
Germanium-068	32	10
Germanium-069	32	10
Germanium-071	32	1000
Germanium-075	32	1000
Germanium-077	32	10
Germanium-078	32	1000
Gold-193	79	100
Gold-194	79	10
Gold-195	79	100
Gold-198	79	100
Gold-198m	79	10
Gold-199	79	100
Gold-200	79	1000
Gold-200m	79	10
Gold-201	79	1000
Hafnium-170	72	100
Hafnium-172	72	1
Hafnium-173	72	100
Hafnium-175	72	100
Hafnium-177m	72	1000
Hafnium-178m	72	0.1
Hafnium-179m	72	100
Hafnium-180m	72	100
Hafnium-181	72	10
Hafnium-182	72	0.1
Hafnium-182m	72	100
Hafnium-183	72	100
Hafnium-184	72	100
Holmium-155	67	1000
Holmium-157	67	1000
Holmium-159	67	1000
Holmium-161	67	1000
Holmium-162	67	1000
Holmium-162m	67	1000
Holmium-164	67	1000
Holmium-164m	67	1000
Holmium-166	67	100
Holmium-166m	67	1
Holmium-167	67	100
Hydrogen-003	1	100
Indium-109	49	100
Indium-110 (4.9 hours)	49	10
Indium-110 (69.1 min)	49	100
Indium-111	49	100
Indium-112	49	1000
Indium-113m	49	1000
Indium-114m	49	10
Indium-115	49	0.1
Indium-115m	49	100
Indium-116m	49	100
Indium-117	49	1000
Indium-117m	49	100
Iodine-119m	53	1000
Iodine-120	53	10
Iodine-120m	53	100
Iodine-121	53	100
Iodine-123	53	10
Iodine-124	53	0.1
Iodine-125	53	0.01
Iodine-126	53	0.01
Iodine-128	53	1000
Iodine-129	53	0.001
Iodine-130	53	1
Iodine-131	53	0.01
Iodine-132	53	10
Iodine-132m	53	10
Iodine-133	53	0.1
Iodine-134	53	100
Iodine-135	53	10
Iridium-182	77	1000
Iridium-184	77	100
Iridium-185	77	100
Iridium-186	77	10
Iridium-187	77	100
Iridium-188	77	10
Iridium-189	77	100
Iridium-190	77	10

Radionuclide Name	Atomic Number	RQ (curies)
Iridium-190m	77	1000
Iridium-192	77	10
Iridium-192m	77	100
Iridium-194	77	100
Iridium-194m	77	10
Iridium-195	77	1000
Iridium-195m	77	100
Iron-052	26	100
Iron-055	26	100
Iron-059	26	10
Iron-060	26	0.1
Krypton-074	36	10
Krypton-076	36	10
Krypton-077	36	10
Krypton-079	36	100
Krypton-081	36	1000
Krypton-083m	36	1000
Krypton-085	36	1000
Krypton-085m	36	100
Krypton-087	36	10
Krypton-088	36	10
Lanthanum-131	57	1000
Lanthanum-132	57	100
Lanthanum-135	57	1000
Lanthanum-137	57	10
Lanthanum-138	57	1
Lanthanum-140	57	10
Lanthanum-141	57	1000
Lanthanum-142	57	100
Lanthanum-143	57	1000
Lead-195m	82	1000
Lead-198	82	100
Lead-199	82	100
Lead-200	82	100
Lead-201	82	100
Lead-202	82	1
Lead-202m	82	10
Lead-203	82	100
Lead-205	82	100
Lead-209	82	1000
Lead-210	82	0.01
Lead-211	82	100
Lead-212	82	10
Lead-214	82	100
Lutetium-169	71	10
Lutetium-170	71	10
Lutetium-171	71	10
Lutetium-172	71	10
Lutetium-173	71	100
Lutetium-174	71	10
Lutetium-174m	71	10
Lutetium-176	71	1
Lutetium-176m	71	1000
Lutetium-177	71	100
Lutetium-177m	71	10
Lutetium-178	71	1000
Lutetium-178m	71	1000
Lutetium-179	71	1000
Magnesium-028	12	10
Manganese-051	25	1000
Manganese-052	25	10
Manganese-053	25	1000
Manganese-054	25	10
Manganese-056	25	100
Manganese-052m	25	1000
Mendelevium-257	101	100
Mendelevium-258	101	1
Mercury-193	80	100
Mercury-193m	80	10
Mercury-194	80	0.1
Mercury-195	80	100
Mercury-195m	80	100
Mercury-197	80	1000
Mercury-197m	80	1000
Mercury-199m	80	1000
Mercury-203	80	10
Molybdenum-090	42	100
Molybdenum-093	42	100
Molybdenum-093m	42	10
Molybdenum-099	42	100
Molybdenum-101	42	1000
Neodymium-136	60	1000
Neodymium-138	60	1000
Neodymium-139	60	1000
Neodymium-139m	60	100
Neodymium-141	60	1000
Neodymium-147	60	10
Neodymium-149	60	100
Neodymium-151	60	1000
Neptunium-232	93	1000
Neptunium-233	93	1000
Neptunium-234	93	10
Neptunium-235	93	1000
Neptunium-236 (1.2 e 5 yr)	93	0.1
Neptunium-236 (22.5 hours)	93	100
Neptunium-237	93	0.01
Neptunium-238	93	10
Neptunium-239	93	100
Neptunium-240	93	100
Nickel-056	28	10

Radionuclide Name	Atomic Number	RQ (curies)
Nickel-057	28	10
Nickel-059	28	100
Nickel-063	28	100
Nickel-065	28	100
Nickel-066	28	10
Niobium-088	41	100
Niobium-089 (122 minutes)	41	100
Niobium-089 (66 minutes)	41	100
Niobium-090	41	10
Niobium-093m	41	100
Niobium-094	41	10
Niobium-095	41	10
Niobium-095m	41	100
Niobium-096	41	10
Niobium-097	41	100
Niobium-098	41	1000
Osmium-180	76	1000
Osmium-181	76	100
Osmium-182	76	100
Osmium-185	76	10
Osmium-189m	76	1000
Osmium-191	76	100
Osmium-191m	76	1000
Osmium-193	76	100
Osmium-194	76	1
Palladium-100	46	100
Palladium-101	46	100
Palladium-103	46	100
Palladium-107	46	100
Palladium-109	46	1000
Phosphorus-032	15	0.1
Phosphorus-033	15	1
Platinum-186	78	100
Platinum-188	78	100
Platinum-189	78	100
Platinum-191	78	100
Platinum-193	78	1000
Platinum-193m	78	100
Platinum-195m	78	100
Platinum-197	78	1000
Platinum-197m	78	1000
Platinum-199	78	1000
Platinum-200	78	100
Plutonium-234	94	1000
Plutonium-235	94	1000
Plutonium-236	94	0.1
Plutonium-237	94	1000
Plutonium-238	94	0.01
Plutonium-239	94	0.01
Plutonium-240	94	0.01
Plutonium-241	94	1
Plutonium-242	94	0.01
Plutonium-243	94	1000
Plutonium-244	94	0.01
Plutonium-245	94	100
Polonium-203	84	100
Polonium-205	84	100
Polonium-207	84	10
Polonium-210	84	0.01
Potassium-040	19	1
Potassium-042	19	100
Potassium-043	19	10
Potassium-044	19	100
Potassium-045	19	1000
Praseodymium-136	59	1000
Praseodymium-137	59	1000
Praseodymium-138m	59	100
Praseodymium-139	59	1000
Praseodymium-142	59	100
Praseodymium-142m	59	1000
Praseodymium-143	59	10
Praseodymium-144	59	1000
Praseodymium-145	59	1000
Praseodymium-147	59	1000
Promethium-141	61	1000
Promethium-143	61	100
Promethium-144	61	10
Promethium-145	61	100
Promethium-146	61	10
Promethium-147	61	10
Promethium-148	61	10
Promethium-148m	61	10
Promethium-149	61	100
Promethium-150	61	100
Promethium-151	61	100
Protactinium-227	91	100
Protactinium-228	91	10
Protactinium-230	91	10
Protactinium-231	91	0.01
Protactinium-232	91	10
Protactinium-233	91	100
Protactinium-234	91	10
Radium-223	88	1
Radium-224	88	10
Radium-225	88	1
Radium-226	88	0.1
Radium-227	88	1000
Radium-228	88	0.1
Radon-220	86	0.1
Radon-222	86	0.1

Radionuclide Name	Atomic Number	RQ (curies)	Radionuclide Name	Atomic Number	RQ (curies)
Rhenium-177	75	1000	Scandium-044	21	100
Rhenium-178	75	1000	Scandium-044m	21	10
Rhenium-181	75	100	Scandium-046	21	10
Rhenium-182 (12.7 hours)	75	10	Scandium-047	21	100
Rhenium-182 (64.0 hours)	75	10	Scandium-048	21	10
Rhenium-184	75	10	Scandium-049	21	1000
Rhenium-184m	75	10	Selenium-070	34	1000
Rhenium-186	75	100	Selenium-073	34	10
Rhenium-186m	75	10	Selenium-073m	34	100
Rhenium-187	75	1000	Selenium-075	34	10
Rhenium-188	75	1000	Selenium-079	34	10
Rhenium-188m	75	1000	Selenium-081	34	1000
Rhenium-189	75	1000	Selenium-081m	34	1000
Rhodium-099	45	10	Selenium-083	34	1000
Rhodium-099m	45	100	Silicon-031	14	1000
Rhodium-100	45	10	Silicon-032	14	1
Rhodium-101	45	10	Silver-102	47	100
Rhodium-101m	45	100	Silver-103	47	1000
Rhodium-102	45	10	Silver-104	47	1000
Rhodium-102m	45	10	Silver-104m	47	1000
Rhodium-103m	45	1000	Silver-105	47	10
Rhodium-105	45	100	Silver-106	47	1000
Rhodium-106m	45	10	Silver-106m	47	10
Rhodium-107	45	1000	Silver-108m	47	10
Rubidium-079	37	1000	Silver-110m	47	10
Rubidium-081	37	100	Silver-111	47	10
Rubidium-081m	37	1000	Silver-112	47	100
Rubidium-082m	37	10	Silver-115	47	1000
Rubidium-083	37	10	Sodium-022	11	10
Rubidium-084	37	10	Sodium-024	11	10
Rubidium-086	37	10	Strontium-080	38	100
Rubidium-087	37	10	Strontium-081	38	1000
Rubidium-088	37	1000	Strontium-083	38	100
Rubidium-089	37	1000	Strontium-085	38	10
Ruthenium-094	44	1000	Strontium-085m	38	1000
Ruthenium-097	44	100	Strontium-087m	38	100
Ruthenium-103	44	10	Strontium-089	38	10
Ruthenium-105	44	100	Strontium-090	38	0.1
Ruthenium-106	44	1	Strontium-091	38	10
Samarium-141	62	1000	Strontium-092	38	100
Samarium-141m	62	1000	Sulfur-035	16	1
Samarium-142	62	1000	Tantalum-172	73	100
Samarium-145	62	100	Tantalum-173	73	100
Samarium-146	62	0.01	Tantalum-174	73	100
Samarium-147	62	0.01	Tantalum-175	73	100
Samarium-151	62	10	Tantalum-176	73	10
Samarium-153	62	100	Tantalum-177	73	1000
Samarium-155	62	1000	Tantalum-178	73	1000
Samarium-156	62	100	Tantalum-179	73	1000
Scandium-043	21	1000	Tantalum-180	73	100

Radionuclide Name	Atomic Number	RQ (curies)
Tantalum-180m	73	1000
Tantalum-182	73	10
Tantalum-182m	73	1000
Tantalum-183	73	100
Tantalum-184	73	10
Tantalum-185	73	1000
Tantalum-186	73	1000
Technetium-093	43	100
Technetium-093m	43	1000
Technetium-094	43	10
Technetium-094m	43	100
Technetium-096	43	10
Technetium-096m	43	1000
Technetium-097	43	100
Technetium-097m	43	100
Technetium-098	43	10
Technetium-099	43	10
Technetium-099m	43	100
Technetium-101	43	1000
Technetium-104	43	1000
Tellurium-116	52	1000
Tellurium-121	52	10
Tellurium-121m	52	10
Tellurium-123	52	10
Tellurium-123m	52	10
Tellurium-125m	52	10
Tellurium-127	52	1000
Tellurium-127m	52	10
Tellurium-129	52	1000
Tellurium-129m	52	10
Tellurium-131	52	1000
Tellurium-131m	52	10
Tellurium-132	52	10
Tellurium-133	52	1000
Tellurium-133m	52	1000
Tellurium-134	52	1000
Terbium-147	65	100
Terbium-149	65	100
Terbium-150	65	100
Terbium-151	65	10
Terbium-153	65	100
Terbium-154	65	10
Terbium-155	65	100
Terbium-156	65	10
Terbium-156m (24.4 hours)	65	1000
Terbium-156m (5.0 hours)	65	1000
Terbium-157	65	100
Terbium-158	65	10
Terbium-160	65	10
Terbium-161	65	100

Radionuclide Name	Atomic Number	RQ (curies)
Thallium-194	81	1000
Thallium-194m	81	100
Thallium-195	81	100
Thallium-197	81	100
Thallium-198	81	10
Thallium-198m	81	100
Thallium-199	81	100
Thallium-200	81	10
Thallium-201	81	1000
Thallium-202	81	10
Thallium-204	81	10
Thorium-226	90	100
Thorium-227	90	1
Thorium-228	90	0.01
Thorium-229	90	0.001
Thorium-230	90	0.01
Thorium-231	90	100
Thorium-232	90	0.001
Thorium-234	90	100
Thulium-162	69	1000
Thulium-166	69	10
Thulium-167	69	100
Thulium-170	69	10
Thulium-171	69	100
Thulium-172	69	100
Thulium-173	69	100
Thulium-175	69	1000
Tin-110	50	100
Tin-111	50	1000
Tin-113	50	10
Tin-117m	50	100
Tin-119m	50	10
Tin-121	50	1000
Tin-121m	50	10
Tin-123	50	10
Tin-123m	50	1000
Tin-125	50	10
Tin-126	50	1
Tin-127	50	100
Tin-128	50	1000
Titanium-044	22	1
Titanium-045	22	1000
Tungsten-176	74	1000
Tungsten-177	74	100
Tungsten-178	74	100
Tungsten-179	74	1000
Tungsten-181	74	100
Tungsten-185	74	10
Tungsten-187	74	100
Tungsten-188	74	10

Radionuclide Name	Atomic Number	RQ (curies)
Uranium-230	92	1
Uranium-231	92	1000
Uranium-232	92	0.01
Uranium-233	92	0.1
Uranium-234	92	0.1
Uranium-235	92	0.1
Uranium-236	92	0.1
Uranium-237	92	100
Uranium-238	92	0.1
Uranium-239	92	1000
Uranium-240	92	1000
Vanadium-047	23	1000
Vanadium-048	23	10
Vanadium-049	23	1000
Xenon-120	54	100
Xenon-121	54	10
Xenon-122	54	100
Xenon-123	54	10
Xenon-125	54	100
Xenon-127	54	100
Xenon-129m	54	1000
Xenon-131m	54	1000
Xenon-133	54	1000
Xenon-133m	54	1000
Xenon-135	54	100
Xenon-135m	54	10
Xenon-138	54	10
Ytterbium-162	70	1000
Ytterbium-166	70	10
Ytterbium-167	70	1000
Ytterbium-169	70	10
Ytterbium-175	70	100
Ytterbium-177	70	1000
Ytterbium-178	70	1000
Yttrium-086	39	10
Yttrium-086m	39	1000
Yttrium-087	39	10
Yttrium-088	39	10
Yttrium-090	39	10
Yttrium-090m	39	100
Yttrium-091	39	10
Yttrium-091m	39	1000
Yttrium-092	39	100
Yttrium-093	39	100
Yttrium-094	39	1000
Yttrium-095	39	1000
Zinc-062	30	100
Zinc-063	30	1000
Zinc-065	30	10
Zinc-069	30	1000
Zinc-069m	30	100
Zinc-071m	30	100
Zinc-072	30	100
Zirconium-086	40	100
Zirconium-088	40	10
Zirconium-089	40	100
Zirconium-093	40	1
Zirconium-095	40	10
Zirconium-097	40	10

NOTES:

m - Signifies a nuclear isomer which is a radionuclide in a higher energy metastable state relative to the parent isotope.

Final RQs for all radionuclides apply to chemical compounds containing the radionuclides and elemental forms regardless of the diameter of pieces of solid material.

An adjusted RQ of one curie applies to all radionuclides not otherwise listed. Whenever the RQs in the SARA Title III Consolidated List and this list are in conflict, the lowest RQ applies.

Notification requirements for releases of mixtures or solutions of radionuclides can be found in 40 CFR section 302.6(b).

THE DESCRIPTIONS OF THE WASTE STREAMS HAVE BEEN TRUNCATED.
THIS LIST SHOULD BE USED FOR REFERENCE ONLY.
COMPLIANCE INFORMATION CAN BE FOUND IN 40 CFR PART 302 AND TABLE 302.4.

RCRA Code	Description	RQ (lbs)
F001	Spent halogenated solvents used in degreasing:	10
	Tetrachloroethylene (CAS No. 127-18-4, RCRA Waste No. U210)	100
	Trichloroethylene (CAS No. 79-01-6, RCRA Waste No. U228)	100
	Methylene chloride (CAS No. 75-09-2, RCRA Waste No. U080)	1000
	1,1,1-Trichloroethane (CAS No. 71-55-6, RCRA Waste No. U226)	1000
	Carbon tetrachloride (CAS No. 56-23-5, RCRA Waste No. U211)	5000
	Chlorinated fluorocarbons	10
F002	Spent halogenated solvents:	10
	Tetrachloroethylene (CAS No. 127-18-4, RCRA Waste No. U210)	100
	Methylene chloride (CAS No. 75-09-2, RCRA Waste No. U080)	1000
	Trichloroethylene (CAS No. 79-01-6, RCRA Waste No. U228)	100
	1,1,1-Trichloroethane (CAS No. 71-55-6, RCRA Waste No. U226)	1000
	Chlorobenzene (CAS No. 108-90-7, RCRA Waste No. U037)	100
	1,1,2-Trichloro-1,2,2-trifluoroethane (CAS No. 76-13-1)	5000
	o-Dichlorobenzene (CAS No. 95-50-1, RCRA Waste No. U070)	100
	Trichlorofluoromethane (CAS No. 75-69-4, RCRA Waste No. U121)	5000
	1,1,2-Trichloroethane (CAS No. 79-00-5, RCRA Waste No. U227)	100
F003	Spent non-halogenated solvents:	100
	Xylene (CAS No. 1330-20-7, RCRA Waste No. U239)	1000
	Acetone (CAS No. 67-64-1, RCRA Waste No. U002)	5000
	Ethyl acetate (CAS No. 141-78-6, RCRA Waste No. U112)	5000
	Ethylbenzene (CAS No. 100-41-4)	1000
	Ethyl ether (CAS No. 60-29-7, RCRA Waste No. U117)	100
	Methyl isobutyl ketone (CAS No. 108-10-1, RCRA Waste No. U161)	5000
	n-Butyl alcohol (CAS No. 71-36-3, RCRA Waste No. U031)	5000
	Cyclohexanone (CAS No. 108-94-1, RCRA Waste No. U057)	5000
	Methanol (CAS No. 67-56-1, RCRA Waste No. U154)	5000
F004	Spent non-halogenated solvents and still bttm. from cresol\nitrobenzene recovery	1000
	Cresols/cresylic acid (CAS No. 1319-77-3, RCRA Waste No. U052)	1000
	Nitrobenzene (CAS No. 98-95-3, RCRA Waste No. U169)	1000
F005	Spent non-halogenated solvents(still bttm.) toluene\methyl ethyl ketone recovery	100
	Toluene (CAS No. 108-88-3, RCRA Waste No. U220)	1000
	Methyl ethyl ketone (CAS No. 78-93-3, RCRA Waste No. U159)	5000
	Carbon disulfide (CAS No. 75-15-0, RCRA Waste No. P022)	100
	Isobutanol (CAS No. 78-83-1, RCRA Waste No. U140)	5000
	Pyridine (CAS No. 110-86-1, RCRA Waste No. U196)	1000
F006	Wastewater treatment sludges from electroplating operations	10
F007	Spent cyanide plating bath solns. from electroplating	10
F008	Plating bath residues from electroplating where cyanides are used	10
F009	Spent stripping/cleaning bath solns. from electroplating where cyanides are used	10
F010	Quenching bath residues from metal heat treating where cyanides are used	10
F011	Spent cyanide soln. from salt bath pot cleaning from metal heat treating	10
F012	Quenching wastewater sludges from metal heat treating where cyanides are used	10
F019	Wastewater treatment sludges from chemical conversion of aluminum coating	10
F020	Wastes from prod. or use of tri/tetrachlorophenol or derivatives	1
F021	Wastes from prod. or use of pentachlorophenol or intermediates	1
F022	Wastes from use of tetra/penta/hexachlorobenzenes	1
F023	Wastes from mat. prod. on equip. which prev. used tri\tetrachlorophenol	1
F024	Wastes from prod. of chlorinated aliphatic hydrocarbons (C1-C5)	1
F025	Lights ends, filters from prod. of chlorinated aliphatic hydrocarbons(C1-C5)	1

THE DESCRIPTIONS OF THE WASTE STREAMS HAVE BEEN TRUNCATED.
THIS LIST SHOULD BE USED FOR REFERENCE ONLY.
COMPLIANCE INFORMATION CAN BE FOUND IN 40 CFR PART 302 AND TABLE 302.4.

RCRA Code	Description	RQ (lbs)
F026	Waste from equipment previously used to prod. tetra/penta/hexachlorobenzenes	1
F027	Discarded wastes containing tetra/penta/hexachlorobenzenes or derivatives	1
F028	Residues from incineration of contaminated soils: F020,F021,F022,F023,F026,F027	1
F032	Wastewaters, process residuals from wood preserving using chlorophenolic solns.	1
F034	Wastewaters, process residuals from wood preserving using creosote formulations	1
F035	Wastewaters, process residuals from wood preserving using arsenic or chromium	1
F037	Petroleum refinery primary oil/water/solids separation sludge	1
F038	Petroleum refinery secondary (emulsified) oil/water/solids separation sludge	1
K001	Wastewater treatment sludge from creosote or pentachlorophenol wood preserving	1
K002	Wastewater treatment sludge from prod. of chrome yellow and orange pigments	1
K003	Wastewater treatment sludge from prod. of molybdate orange pigments	1
K004	Wastewater treatment sludge from prod. of zinc yellow pigments	10
K005	Wastewater treatment sludge from prod. of chrome green pigments	1
K006	Wastewater treatment sludge from prod. of chrome oxide green pigments anyhydrous	10
K007	Wastewater treatment sludge from prod. of iron blue pigments	10
K008	Oven residue from prod. of chrome oxide green pigments	10
K009	Dist. bottoms from prod. of acetaldehyde from ethylene	10
K010	Dist. side cuts from prod. of acetaldehyde from ethylene	10
K011	Bottom stream from wastewater stripper in acrylonitrile prod.	10
K013	Bottom stream from acetonitrile column in acrylonitrile prod.	10
K014	Bottoms from acetonitrile purification column in acrylonitrile prod.	5000
K015	Still bottoms from the dist. of benzyl chloride	10
K016	Heavy ends or dist. residues from prod. of carbon tetrachloride	1
K017	Heavy ends from the purification column in epichlorohydrin prod.	10
K018	Heavy ends from the fractionation column in ethyl chloride prod.	1
K019	Heavy ends from the dist. of ethylene dichloride during its prod.	1
K020	Heavy ends from the dist. of vinyl chloride during prod. of the monomer	1
K021	Aqueous spent antimony catalyst waste from fluoromethanes prod.	10
K022	Dist. bottom tars from prod. of phenol/acetone from cumene	1
K023	Dist. light ends from prod. of phthalic anhydride from naphthalene	5000
K024	Dist. bottoms from prod. of phthalic anhydride from naphthalene	5000
K025	Dist. bottoms from prod. of nitrobenzene by nitration of benzene	10
K026	Stripping still tails from the prod. of methyl ethyl pyridines	1000
K027	Centrifuge/dist. residues from toluene diisocyanate prod.	10
K028	Spent catalyst from hydrochlorinator reactor in prod. of 1,1,1-trichloroethane	1
K029	Waste from product steam stripper in prod. of 1,1,1-trichloroethane	1
K030	Column bottoms(heavy ends) from prod. of trichloroethylene and perchloroethylene	1
K031	By-product salts generated in the prod. msma and cacodylic acid	1
K032	Wastewater treatment sludge from the prod. of chlordane	10
K033	Wastewaster/scrubwater from chlorination during prod. of chlordane	10
K034	Filter solids from filtration of hexachlorocyclopentadiene in chlordane prod.	10
K035	Wastewater treatment sludges from the prod. of creosote	1
K036	Still bottoms from toluene reclamation distillation in disulfoton prod.	1
K037	Wastewater treatment sludges from the prod. of disulfoton	1
K038	Wastewater from the washing and stripping of phorate production	10
K039	Filter cake from filtration during prod. of phorate	10
K040	Wastewater treatment sludge from the prod. of phorate	10
K041	Wastewater treatment sludge from the prod. of toxaphene	1
K042	Heavy ends from dist. of tetrachlorobenzene in the prod. of 2,4,5-T	10
K043	2,6-Dichlorophenol waste from the prod. of 2,4-D	10

THE DESCRIPTIONS OF THE WASTE STREAMS HAVE BEEN TRUNCATED.
THIS LIST SHOULD BE USED FOR REFERENCE ONLY.
COMPLIANCE INFORMATION CAN BE FOUND IN 40 CFR PART 302 AND TABLE 302.4.

RCRA Code	Description	RQ (lbs)
K044	Wastewater treatment sludge from manuf. and processing of explosives	10
K045	Spent carbon from treatment of wastewater containing explosives	10
K046	Wastewater sludge from manuf.,formulating,loading of lead-based initiating compd	100
K047	Pink/red water from TNT operations	10
K048	Dissolved air flotation (DAF) float from the petroleum refining industry	1
K049	Slop oil emulsion solids from the petroleum refining industry	1
K050	Heat exchanger bundle cleaning sludge from petroleum refining industry	10
K051	API separator sludge from the petroleum refining industry	1
K052	Tank bottoms (leaded) from the petroleum refining industry	10
K060	Ammonia still lime sludge from coking operations	1
K061	Emission control dust/sludge from primary prod. of steel in electric furnaces	1
K062	Spent pickle liquor generated by steel finishing: (SIC codes 331 and 332)	1
K064	Acid plant blowdown sludge from blowdown slurry from primary copper prod.	1
K065	Surface impoundment solids at primary lead smelting facilities	1
K066	Sludge from treatment of wastewater(acid plant blowdown) from primary zinc prod.	1
K069	Emission control dust/sludge from secondary lead smelting	1
K071	Brine purification muds from mercury cell process in chlorine production	1
K073	Chlorinated hydrocarbon waste in chlorine production	10
K083	Distillation bottoms from aniline extraction	100
K084	Wastewater sludges from prod. of veterinary pharm. from arsenic compds.	1
K085	Distillation or fractionation column bottoms in prod. of chlorobenzenes	10
K086	Wastes/sludges from prod. of inks from chromium and lead compds.	1
K087	Decanter tank tar sludge from coking operations	100
K088	Spent potliners from primary aluminum reduction	1
K090	Emission control dust/sludge from ferrochromiumsilicon prod.	1
K091	Emission control dust/sludge from ferrochromium prod.	1
K093	Dist. light ends from prod. of phthalic anhydride by ortho-xylene	5000
K094	Dist. bottoms in prod. of phthalic anhydride by ortho-xylene	5000
K095	Distillation bottoms in prod. of 1,1,1-trichloroethane	100
K096	Heavy ends from dist. column in prod. of 1,1,1-trichloroethane	100
K097	Vacuum stripper discharge from the chlordane chlorinator in prod. of chlordane	1
K098	Untreated process wastewater from the prod. of toxaphene	1
K099	Untreated wastewater from the prod. of 2,4-D	10
K100	Waste leaching soln. from acid leaching of emission dust in 2nd lead smelting	1
K101	Dist. tar residue from aniline in prod. of veterinary pharm. from arsenic compd.	1
K102	Residue from activated carbon in prod. of veterinary pharm. from arsenic compds.	1
K103	Process residues from aniline extraction from the prod. of aniline	100
K104	Combined wastewater streams generated from prod. of nitrobenzene/aniline	10
K105	Aqueous stream from washing in prod. of chlorobenzenes	10
K106	Wastewater treatment sludge from mercury cell process in chlorine prod.	1
K107	Column bottoms from separation in prod. of UDMH from carboxylic acid hydrazides	10
K108	Condensed column overheads and vent gas from prod. of UDMH from -COOH hydrazides	10
K109	Spent filter catridges from purif. of UDMH prod. from carboxylic acid hydrazides	10
K110	Condensed column overheads from prod. of UDMH from carboxylic acid hydrazides	10
K111	Product washwaters from prod. of dinitrotoluene via nitration of benzene	10
K112	Reaction by-product water from drying of toluenediamine during its prod.	10
K113	Condensed liquid light ends from purification of toluenediamine during its prod.	10
K114	Vicinals from purification of toluenediamine during its prod.	10
K115	Heavy ends from purification of toluenediamine during its prod.	10
K116	Organic condensate solvent recovery system in prod. of toluene diisocyanate	10

THE DESCRIPTIONS OF THE WASTE STREAMS HAVE BEEN TRUNCATED.
THIS LIST SHOULD BE USED FOR REFERENCE ONLY.
COMPLIANCE INFORMATION CAN BE FOUND IN 40 CFR PART 302 AND TABLE 302.4.

RCRA Code	Description	RQ (lbs)
K117	Wastewater from vent gas scrubber in prod. of ethylene bromide prod. from ethene	1
K118	Spent absorbant solids in purification of ethylene dibromide manuf. from ethene	1
K123	Process waterwater from the prod. of ethylenebisdithiocarbamic acid and salts	10
K124	Reactor vent scubber water from prod of ethylenebisdithiocarbamic acid and salts	10
K125	Solids formed in the prod. of ethylenebisdithiocarbamic acid and salts	10
K126	Dust/sweepings from the prod. of ethylenebisdithiocarbamic acid and salts	10
K131	Wastewater and spent sulfuric acid from the prod. of methyl bromide	100
K132	Spent absorbent and waste water from the prod. of methyl bromide	1000
K136	Still bottoms from purification of ethylene dibromide manuf. from ethene	1
D001	Unlisted hazardous wastes characteristic of ignitability	100
D002	Unlisted hazardous wastes characteristic of corrosivity	100
D003	Unlisted hazardous wastes characteristic of reactivity	100
	Unlisted hazardous wastes characteristic of toxicity:	
D004	Arsenic	1
D005	Barium	1000
D018	Benzene	10
D006	Cadmium	10
D019	Carbon tetrachloride	10
D020	Chlordane	1
D021	Chlorobenzene	100
D022	Chloroform	10
D007	Chromium	10
D023	o-Cresol	1000
D024	m-Cresol	1000
D026	Cresol	1000
D025	p-Cresol	1000
D016	2,4-D	100
D027	1,4-Dichlorobenzene	100
D028	1,2-Dichloroethane	100
D029	1,1-Dichloroethylene	100
D030	2,4-Dinitrotoluene	10
D012	Endrin	1
D031	Heptachlor (and epoxide)	1
D032	Hexachlorobenzene	10
D033	Hexachlorobutadiene	1
D034	Hexachloroethane	100
D008	Lead	1
D013	Lindane	1
D009	Mercury	1
D014	Methoxychlor	1
D035	Methyl ethyl ketone	5000
D036	Nitrobenzene	1000
D037	Pentachlorophenol	10
D038	Pyridine	1000
D010	Selenium	10
D011	Silver	1
D039	Tetrachloroethylene	100
D015	Toxaphene	1
D017	2,4,5-TP	100
D040	Trichloroethylene	100

THE DESCRIPTIONS OF THE WASTE STREAMS HAVE BEEN TRUNCATED.
THIS LIST SHOULD BE USED FOR REFERENCE ONLY.
COMPLIANCE INFORMATION CAN BE FOUND IN 40 CFR PART 302 AND TABLE 302.4.

RCRA Code	Description	RQ (lbs)
D041	2,4,5-Trichlorophenol	10
D042	2,4,6-Trichlorophenol	10
D043	Vinyl chloride	100

☆ U.S. GOVERNMENT PRINTING OFFICE:1992-617-003/47087

Sample LEPC Questionnaire

Superfund Amendments and Reauthorization Act
Title III Section 302 Facility Questionnaire
(Revised September, 1990)

Any facility that has reported greater than the threshold planning quantity (TPQ) of any "extremely hazardous substance" covered by SARA Title III must cooperate in the preparation of the community emergency plan for hazardous materials. This questionnaire is an initial step in the development of a cooperative planning process involving your facility, local fire department, emergency manager, and regional review committee.

If you require assistance in completing the questionnaire or desire an on-site interview, please contact ____________________, phone ______________.

I. FACILITY INFORMATION

1. Facility Name: ____

2. Type of Facility (check):

   [] Wholesale trade   [] Retail trade   [] Manufacturing
   [] Agriculture   [] Electric, Gas, Sanitary Services
   [] Education   [] Other________________________________

3. ERC Identification Number: ______________________________

4. Street Address: ____

   ____

5. Facility Emergency Coordinator, and Alternate (Facilities which have a 24 hr. phone number for after hours contact with personnel may list that number in lieu of seperate work and home numbers.)

   a. Coordinator Name: ______________________________________

      Work Address:____

      ____

      Home Address:____

      ____

      Work Phone:_____________   Home Phone :____________

   b. Alternate Name: ______________________________________

      Work Address:____

      ____

      Home Address:____

      ____

      Work Phone: ____________   Home Phone: ____________

6. How many people typically work at or occupy your facility?

____

7. Does your facility operate on a 24-hour basis?

_____ Yes _____ No

8. Describe any unusual geographic features or sensitive environmental areas near or at your facility that might affect or alter an emergency response.

____
____
____
____
____
____

II. HAZARDOUS MATERIALS INFORMATION

1. Please list the Extremely Hazardous Substances (EHS) Section 302 chemicals that your facility has on-site. (A copy of your facility's most recent Tier II report identifying the extremely hazardous substances on site may be attached in lieu of completing this section.)

EHS Chemical #1-Name: ________________ CAS Number: ____________

Check all that apply:
[] Solid [] Liquid [] Gas [] Pure [] In Mixture

Where is chemical used and stored? (You may attach a site plan or leave blank if confidential)

____
____
____

EHS Chemical #2-Name: ________________ CAS Number: ____________

Check all that apply:
[] Solid [] Liquid [] Gas [] Pure [] In Mixture

Where is chemical used and stored? (You may attach a site plan or leave blank if confidential)

____
____
____

EHS Chemical #3-Name: ________________ CAS Number: ____________

Check all that apply:
[] Solid [] Liquid [] Gas [] Pure [] In Mixture

Where is chemical used and stored? (You may attach a site plan or leave blank if confidential)

____
____
____

EHS Chemical #4-Name: ________________ CAS Number: ____________

Check all that apply:
[] Solid [] Liquid [] Gas [] Pure [] In Mixture

Where is chemical used and stored? (You may attach a site plan or leave blank if confidential)

____
____
____

EHS Chemical #5-Name: ________________ CAS Number: ____________

Check all that apply:
[] Solid [] Liquid [] Gas [] Pure [] In Mixture

Where is chemical used and stored? (You may attach a site plan or leave blank if confidential)

____
____
____

2. For each of the Extremely Hazardous Substances (EHS) listed on the previous page, please provide the following information.

	#1	#2	#3	#4	#5
a. Largest individual shipment of EHS chemical or its mixture (pounds, gallons or cubic ft)	______	______	______	______	______
b. Largest container size or group of interconnected containers of EHS chemical or its mixture (pounds, gallons or cubic ft)	______	______	______	______	______
c. If mixture reported in a and b of above, indicate from material safety data sheet weight percentage of EHS chemical	______%	______%	______%	______%	______%
d. Maximum amount of EHS chemical stored (pounds, gallons or cubic ft)	______	______	______	______	______
e. If at your facility for only part of year, indicate which months	______	______	______	______	______

3. Are any containers located in a diked area?

_____ Yes _____ No

4. What transportation method(s) are used for shipment of each EHS?

	#1	#2	#3	#4	#5
a. Road	______	______	______	______	______
Cargo Tank	______	______	______	______	______
Semi Van	______	______	______	______	______
Straight Van	______	______	______	______	______
b. Air	______	______	______	______	______
c. Rail	______	______	______	______	______
d. Pipe Line	______	______	______	______	______
e. Water	______	______	______	______	______

5. What are the <u>primary routes</u> used to transport extremely hazardous substances to your facility?

____
____
____
____
____

6. What are the <u>secondary routes</u> used to transport extremely hazardous substances to your facility?

____
____
____
____
____

7. Has a hazard analysis been prepared for your facility which would define potential populations affected by a chemical emergency?

_____ Yes _____ No

III. <u>METHODS AND PROCEDURES TO RESPOND TO EMERGENCIES</u>

1. Should an accident release of hazardous materials occur at your facility, how would such a release to detected? (i.e. employee observation, special monitoring equipment)

____
____
____
____
____

2. a. Does your facility have a written Emergency Plan that covers hazardous materials response?

_____ Yes _____ No

b. If no, would you like assistance in preparing an Emergency Plan?

_____ Yes _____ No

3. Does your facility have a plan for the evacuation of employees and other occupants?

_____ Yes _____ No

4. Is on-site emergency response equipment available to provide on-site initial response efforts?

_____ Yes _____ No

5. a. Are on-site trained personnel available to provide on-site initial response efforts?

_____ Yes _____ No

b. If yes, please briefly describe type and level of training.

____
____
____
____
____

6. Do you have written preplanned Emergency Protocols established with:

a. Fire _____ Yes _____ No
b. Police _____ Yes _____ No
c. Emergency Medical Services (ex: ambulance, etc.) _____ Yes _____ No
d. Hospital(s) _____ Yes _____ No
e. Utility (ex: sewers, gas, etc.)
f. Clean-up contractors _____ Yes _____ No
g. Mutual aid with other facilities _____ Yes _____ No
h. Other (List) _______________________________
____
____
____

7. a. Does your facility have the capability and plans for responding to off-site emergencies?

_____ Yes _____ No

b. If yes, is this limited to the company's products?

_____ Yes _____ No

IV. NOTIFICATION

1. Describe alert and warning procedures and equipment.

   a. On-Site:

   ____
   ____
   ____

   b. Off-Site:

   ____
   ____
   ____

2. Have you worked with your local community in developing these procedures?

   _____ Yes _____ No

3. Do you provide training exercises on emergency procedures for:

Employees	_____ Yes	_____ No
Local First Responders	_____ Yes	_____ No
Public	_____ Yes	_____ No

V. DESCRIPTION OF EMERGENCY EQUIPMENT/RESOURCES

1. What chemical emergency monitoring equipment do you have?

   weather equipment______________________
   pH meters (indicate fixed or portable)______________________
   combustible gas indicator______________________
   colormetric indicator tubes (i.e. draeger tubes)______________________
   radiation detector______________________
   chlorine kits (A.B.C.)______________________
   oxygen concentration meter______________________
   organic vapor monitor______________________
   other:______________________
   ____

2. What personnel protective equipment do you have (how many)?

   positive pressure respirators______________________
   self-contained breathing apparatus (SCBA)______________________
   SCBA tanks (duration)______________________
   mobile cascade______________________
   cascade with compressor______________________
   fully encapsulated suits (indicate type)______________________
   full protective turnout gear______________________
   boots and gloves______________________
   helmets with eye protection______________________
   other:______________________
   ____

3. Which of the following equipment/supplies do you have (quantity)?

foam (indicate type)__________
sand__________
other absorbents (indicate type)__________
fire equipment:
    pumper__________
    ladder truck__________
    tanker__________
    rescue squad__________
    EMT__________
    paramedic__________
    other:__________
off-road vehicles__________
communication vehicle__________
multipurpose vehicles__________
other:__________

4. Other Equipment/Supplies__________

5. Would you be willing to share your equipment/supplies on a reimbursable basis?

_____ Yes   _____ No

6. Identify additional professional/technical resources that may be called upon by the facility to support regular staff in the event of an incident:

Name	Organization	Phone: Home/Work	Specialty

Please return the questionnaire to:

TCLP-Toxic Contaminants

TABLE 1—MAXIMUM CONCENTRATION OF CONTAMINANTS FOR THE TOXICITY CHARACTERISTIC

EPA HW No.[1]	Contaminant	CAS No.[2]	Regulatory Level (mg/L)
D004	Arsenic	7440-38-2	5.0
D005	Barium	7440-39-3	100.0
D018	Benzene	71-43-2	0.5
D006	Cadmium	7440-43-9	1.0
D019	Carbon tetrachloride	56-23-5	0.5
D020	Chlordane	57-74-9	0.03
D021	Chlorobenzene	108-90-7	100.0
D022	Chloroform	67-66-3	6.0
D007	Chromium	7440-47-3	5.0.
D023	o-Cresol	95-48-7	[4] 200.0
D024	m-Cresol	108-39-4	[4] 200.0
D025	p-Cresol	106-44-5	[4] 200.0
D026	Cresol		[4] 200.0
D016	2,4-D	94-75-7	10.0
D027	1,4-Dichlorobenzene	106-46-7	7.5
D028	1,2-Dichloroethane	107-06-2	0.5
D029	1,1-Dichloroethylene	75-35-4	0.7
D030	2,4-Dinitrotoluene	121-14-2	[3] 0.13
D012	Endrin	72-20-8	0.02
D031	Heptachlor (and its epoxide).	76-44-8	0.008
D032	Hexachlorobenzene	118-74-1	[3] 0.13
D033	Hexachlorobutadiene	87-68-3	0.5
D034	Hexachloroethane	67-72-1	3.0
D008	Lead	7439-92-1	5.0
D013	Lindane	58-89-9	0.4
D009	Mercury	7439-97-6	0.2
D014	Methoxychlor	72-43-5	10.0
D035	Methyl ethyl ketone	78-93-3	200.0
D036	Nitrobenzene	98-95-3	2.0
D037	Pentrachlorophenol	87-86-5	100.0
D038	Pyridine	110-86-1	[3] 5.0
D010	Selenium	7782-49-2	1.0
D011	Silver	7440-22-4	5.0
D039	Tetrachloroethylene	127-18-4	0.7
D015	Toxaphene	8001-35-2	0.5
D040	Trichloroethylene	79-01-6	0.5
D041	2,4,5-Trichlorophenol	95-95-4	400.0
D042	2,4,6-Trichlorophenol	88-06-2	2.0
D017	2,4,5-TP (Silvex)	93-72-1	1.0
D043	Vinyl chloride	75-01-4	0.2

[1] Hazardous waste number.

[2] Chemical abstracts service number.

[3] Quantitation limit is greater than the calculated regulatory level. The quantitation limit therefore becomes the regulatory level.

[4] If o-, m-, and p-Cresol concentrations cannot be differentiated, the total cresol (D026) concentration is used. The regulatory level of total cresol is 200 mg/l.

EPA Regional Offices

EPA Regional Offices

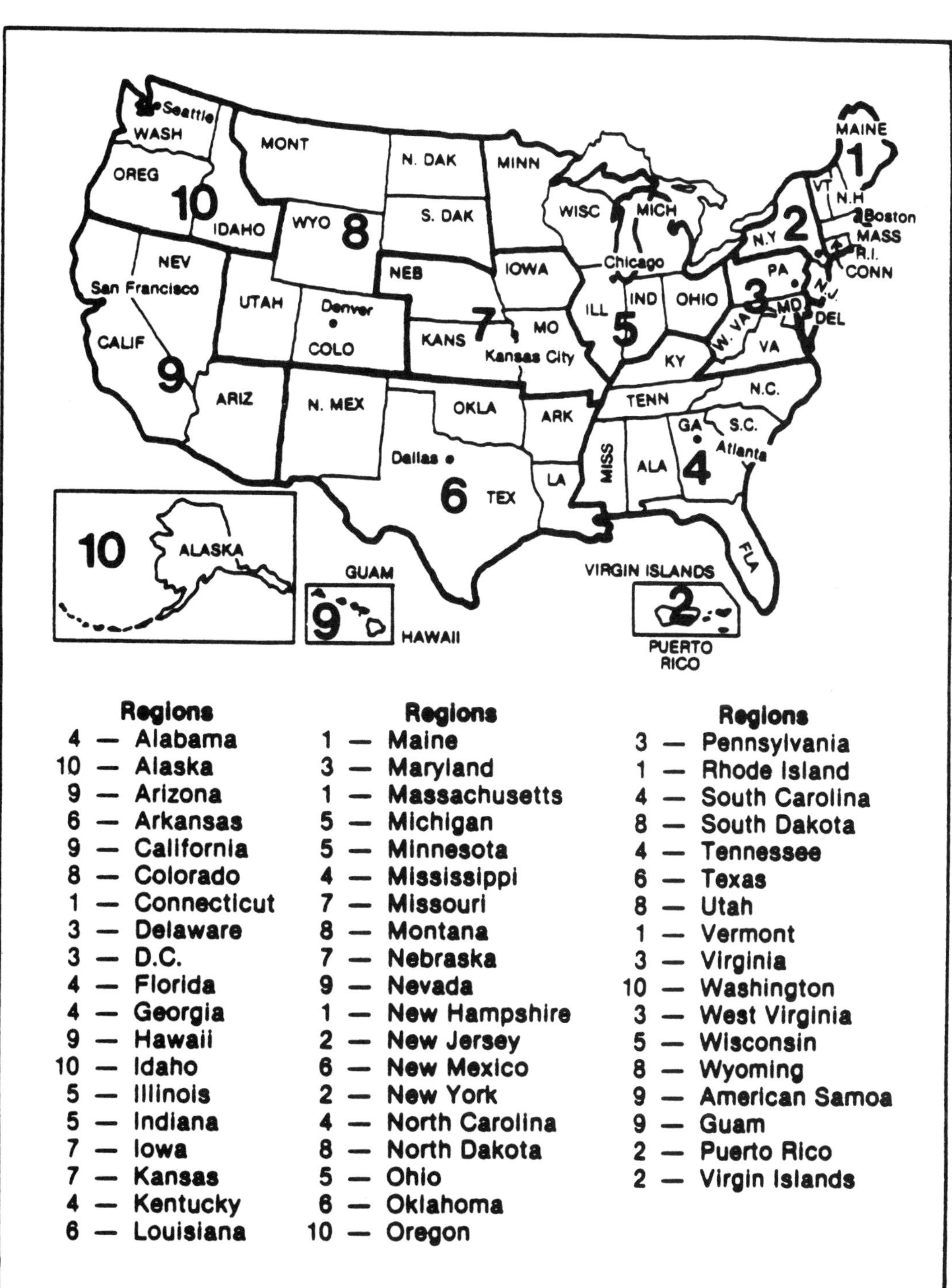

Regions		Regions		Regions	
4	Alabama	1	Maine	3	Pennsylvania
10	Alaska	3	Maryland	1	Rhode Island
9	Arizona	1	Massachusetts	4	South Carolina
6	Arkansas	5	Michigan	8	South Dakota
9	California	5	Minnesota	4	Tennessee
8	Colorado	4	Mississippi	6	Texas
1	Connecticut	7	Missouri	8	Utah
3	Delaware	8	Montana	1	Vermont
3	D.C.	7	Nebraska	3	Virginia
4	Florida	9	Nevada	10	Washington
4	Georgia	1	New Hampshire	3	West Virginia
9	Hawaii	2	New Jersey	5	Wisconsin
10	Idaho	6	New Mexico	8	Wyoming
5	Illinois	2	New York	9	American Samoa
5	Indiana	4	North Carolina	9	Guam
7	Iowa	8	North Dakota	2	Puerto Rico
7	Kansas	5	Ohio	2	Virgin Islands
4	Kentucky	6	Oklahoma		
6	Louisiana	10	Oregon		

REGION 1

Connecticut
Maine
Massachusetts
New Hampshire
Rhode Island
Vermont

REGION 2

New Jersey
New York
Puerto Rico
Virgin Islands

REGION 3

Delaware
District of Columbia
Maryland
Pennsylvania
Virginia
West Virginia

REGION 4

Alabama
Florida
Georgia
Kentucky
Mississippi
North Carolina
South Carolina
Tennessee

REGION 5

Illinois
Indiana
Michigan
Minnesota
Ohio
Wisconsin

REGION 6

Arkansas
Louisiana
New Mexico
Oklahoma
Texas

REGION 7

Iowa
Kansas
Missouri
Nebraska

REGION 8

Colorado
Montana
North Dakota
South Dakota
Utah
Wyoming

REGION 9

Arizona
California
Hawaii
Nevada
American Samoa
Guam

REGION 10

Alaska
Idaho
Oregon
Washington

Regional Directory Contacts

Program Office	Directory Contact
Region 1	USEPA Region 1 Sandy Warner PIM-91 John F. Kennedy Federal Building One Congress Street Boston, MA 02203 FTS : 8-835-3371 FAX : 8-835-3736
Region 2	USEPA Region 2 Sally Kaufman Jacob K. Javitz Federal Building 26 Federal Plaza Room 937 New York, NY 10278 FTS : 8-264-5852 FAX : 8-264-8100
Region 3	USEPA Region 3 Bob Braster 3PM51 841 Chestnut Building Philadelphia, PA 19107 FTS : 8-597-9855 FAX : 8-597-7906
Region 4	USEPA Region 4 Drunell Williams 345 Courtland Street, NE Atlanta, GA 30365 FTS : 8-257-2316 FAX : 8-257-4486 Alternate: Rick Shekell FTS : 8-257-2316
Region 5	USEPA Region 5 Stephen Goranson 230 South Dearborn Street MAIL CODE 5MI-16 Chicago, IL 60604 FTS : 8-353-2074 FAX : 8-886-1515 Alternate: Arthur Kawatachi FTS : 8-353-2074

Program Office	**Directory Contact**
Region 6	USEPA Region 6 Wanda Poole 6M-AO First Interstate Bank Tower at Fountain Place 1445 Ross Avenue Dallas, TX 75202-2733 FTS : 8-255-6570 FAX : 8-255-2142
Region 7	USEPA Region 7 Diann Collier 726 Minnesota Avenue Kansas City, KS 66101 FTS : 8-276-7222 FAX : 8-276-7467 <u>Alternate:</u> Lorri Estes FTS : 8-276-7886 FAX : 8-276-7467
Region 8	USEPA Region 8 Mel McCottry Bill Ross 999 18th Street Suite 500 Denver, CO 80202 FTS : 8-330-1645 FAX : 8-330-1647
Region 9	USEPA Region 9 Paul Helliker P-7-1 75 Hawthorne Street San Francisco, CA 94105 FTS : 8-484-1701 FAX : 8-484-1680
Region 10	USEPA Region 10 Jim Peterson 1200 6th Street Seattle, WA 98101 FTS : 8-399-2977 FAX : 8-399-4672

NOTE:

These regional contacts provided the data for the Regional Directory. Please coordinate updates to this information with these regional contacts.

Sample Material Safety Data Sheet (MSDS)

Material Safety Data Sheet
May be used to comply with
OSHA's Hazard Communication Standard,
29 CFR 1910.1200. Standard must be
consulted for specific requirements.

U.S. Department of Labor
Occupational Safety and Health Administration
(Non-Mandatory Form)
Form Approved
OMB No. 1218-0072

IDENTITY *(As Used on Label and List)*	*Note: Blank spaces are not permitted. If any item is not applicable, or no information is available, the space must be marked to indicate that.*

Section I

Manufacturer's Name	Emergency Telephone Number
Address *(Number, Street, City, State, and ZIP Code)*	Telephone Number for Information
	Date Prepared
	Signature of Preparer *(optional)*

Section II — Hazardous Ingredients/Identity Information

Hazardous Components (Specific Chemical Identity; Common Name(s))	OSHA PEL	ACGIH TLV	Other Limits Recommended	% *(optional)*

Section III — Physical/Chemical Characteristics

Boiling Point		Specific Gravity (H_2O = 1)	
Vapor Pressure (mm Hg.)		Melting Point	
Vapor Density (AIR = 1)		Evaporation Rate (Butyl Acetate = 1)	
Solubility in Water			
Appearance and Odor			

Section IV — Fire and Explosion Hazard Data

Flash Point (Method Used)	Flammable Limits	LEL	UEL
Extinguishing Media			
Special Fire Fighting Procedures			
Unusual Fire and Explosion Hazards			

(Reproduce locally)

OSHA 174, Sept. 1985

Section V — Reactivity Data

Stability	Unstable		Conditions to Avoid
	Stable		

Incompatibility (*Materials to Avoid*)

Hazardous Decomposition or Byproducts

Hazardous Polymerization	May Occur		Conditions to Avoid
	Will Not Occur		

Section VI — Health Hazard Data

Route(s) of Entry: Inhalation? Skin? Ingestion?

Health Hazards (*Acute and Chronic*)

Carcinogenicity: NTP? IARC Monographs? OSHA Regulated?

Signs and Symptoms of Exposure

Medical Conditions
Generally Aggravated by Exposure

Emergency and First Aid Procedures

Section VII — Precautions for Safe Handling and Use

Steps to Be Taken in Case Material Is Released or Spilled

Waste Disposal Method

Precautions to Be Taken in Handling and Storing

Other Precautions

Section VIII — Control Measures

Respiratory Protection (*Specify Type*)

Ventilation	Local Exhaust	Special
	Mechanical (*General*)	Other

Protective Gloves	Eye Protection

Other Protective Clothing or Equipment

Work/Hygienic Practices

☆ U.S.G.P.O.: 1986-491-529/45775

Sample Tier I Form

Revised June 1990

Page ____ of ______ pages
Form Approved OMB No. 2050–0072

Tier One EMERGENCY AND HAZARDOUS CHEMICAL INVENTORY

Aggregate Information by Hazard Type

FOR OFFICIAL USE ONLY — ID # ______ Date Received ______

Important: Read instructions before completing form

Reporting Period From January 1 to December 31, 19______

Facility Identification

Name ______
Street ______
City ______ County ______ State ______ Zip ______

SIC Code [| | |] Dun & Brad Number [|]-[| |]-[| | |]

Owner/Operator

Name ______
Mail Address ______
Phone ( ) ______

Emergency Contacts

Name ______
Title ______
Phone ( ) ______
24 Hour Phone ( ) ______

Name ______
Title ______
Phone ( ) ______
24 Hour Phone ( ) ______

☐ Check if information below is identical to the information submitted last year.

☐ Check if site plan is attached

Physical Hazards

Hazard Type	Max Amount*	Average Daily Amount*	Number of Days On-Site	General Location
Fire	[|]	[|]	[| |]	______
Sudden Release of Pressure	[|]	[|]	[| |]	______
Reactivity	[|]	[|]	[| |]	______

Health Hazards

Hazard Type	Max Amount*	Average Daily Amount*	Number of Days On-Site	General Location
Immediate (acute)	[|]	[|]	[| |]	______
Delayed (Chronic)	[|]	[|]	[| |]	______

Certification *(Read and sign after completing all sections)*

I certify under penalty of law that I have personally examined and am familiar with the information submitted in pages one through ________, and that based on my inquiry of those individuals responsible for obtaining the information, I believe that the submitted information is true, accurate and complete.

______
Name and official title of owner/operator OR owner/operator's authorized representative

______ Signature ______ Date signed

* Reporting Ranges

Range Code	Weight Range in Pounds From...	To...
01	0	99
02	100	999
03	1000	9,999
04	10,000	99,999
05	100,000	999,999
06	1,000,000	9,999,999
07	10,000,000	49,999,999
08	50,000,000	99,999,999
09	100,000,000	499,999,999
10	500,000,000	999,999,999
11	1 billion	higher than 1 billion

TIER ONE INSTRUCTIONS

GENERAL INFORMATION

Submission of this form is required by Title III of the Superfund Amendments and Reauthorization Act of 1986, Title III Section 312, Public Law 99-499, codified at 42 U.S.C. Section 11022.

CERTIFICATION
The owner or operator or the officially designated representative of the owner or operator must certify that all information included in the Tier I submission is true, accurate, and complete. On the Tier I form, enter your full name and official title. Sign your name and enter the current date. Also, enter the total number of pages in the submission, including all attachments.

The purpose of this form is to provide State and local officials and the public with information on the general types and locations of hazardous chemicals present at your facility during the past year.

YOU MUST PROVIDE ALL INFORMATION REQUESTED ON THIS FORM.

You may substitute the Tier Two form for this Tier One form. (The Tier Two form provides detailed information and must be submitted in response to a specific request from State or local officials.)

WHO MUST SUBMIT THIS FORM

Section 312 of Title III requires that the owner or operator of a facility submit this form if, under regulations implementing the Occupational Safety and Health Act of 1970, the owner or operator is required to prepare or have available Material Safety Data Sheets (MSDS) for hazardous chemicals present at the facility. MSDS requirements are specified in the Occupational Safety and Health Administration (OSHA) Hazard Communication Standard, found in Title 29 of the Code of Federal Regulations at §1910.1200.

This form does not have to be submitted if all of the chemicals located at your facility are excluded under Section 311(e) of Title III or if the weight of each covered hazardous chemical never equals or exceeds the minimum threshold listed in Title III Section 312 during the reporting year.

WHAT CHEMICALS ARE INCLUDED

You must report the information required on this form for every hazardous chemical for which you are required to prepare or have available an MSDS under the Hazard Communication Standard, unless the chemicals are excluded under Section 311(e) of Title III or they are below the minimum reporting thresholds.

WHAT CHEMICALS ARE EXCLUDED

Section 311(e) of Title III excludes the following substances:

(i) Any food, food additive, color additive, drug, or cosmetic regulated by the Food and Drug Administration;

(ii) Any substance present as a solid in any manufactured item to the extent exposure to the substance does not occur under normal conditions of use;

(iii) Any substance to the extent it is used for personal, family, or household purposes, or is present in the same form and concentration as a product packaged for distribution and use by the general public;

(iv) Any substance to the extent it is used in a research laboratory or a hospital or other medical facility under the direct supervision of a technically qualified individual;

(v) Any substance to the extent it is used in routine agricultural operations or is a fertilizer held for sale by a retailer to the ultimate customer.

OSHA regulations, Section 1910.1200(b), stipulate exemptions from the requirement to prepare or have available an MSDS.

REPORTING THRESHOLDS

Minimum thresholds have been established for Tier One/Tier Two reporting under Title III, Section 312. These thresholds are as follows:

For Extremely Hazardous Substances (EHSs) designated under section 302 of Title III, the reporting threshold is 500 pounds (or 227 kg.) or the threshold planning quantity (TPQ), whichever is lower;

For all other hazardous chemicals for which facilities are required to have or prepare an MSDS, the minimum reporting threshold is 10,000 pounds (or 4,540 kg.).

You need to report hazardous chemicals that were present at your facility at any time during the previous calendar year at levels that equal or exceed these thresholds. For instructions on threshold determinations for components of mixtures, see "What About Mixtures?" on page 3 of these instructions.

WHEN TO SUBMIT THIS FORM

Owners or operators of facilities that have hazardous chemicals on hand in quantities equal to or greater than set threshold levels must submit either Tier One or Tier Two Forms by March 1.

WHERE TO SUBMIT THIS FORM

Send one completed inventory form to each of the following organizations:

1. Your State emergency response commission
2. Your local emergency planning committee
3. The fire department with jurisdiction over your facility.

PENALTIES

Any owner or operator of a facility who fails to submit or supplies false Tier One information shall be liable to the United States for a civil penalty of up to $25,000 for each such violation. Each day a violation continues shall constitute a separate violation. In addition, any citizen may commence a civil action on his or her own behalf against any owner or operator who fails to submit Tier One information.

INSTRUCTIONS

Please read these instructions carefully. Print or type all responses.

You may use the Tier Two form as a worksheet for completing Tier One. Filling in the Tier Two chemical information section should help you assemble your Tier One responses.

If your responses require more than one page, fill in the page number at the top of the form.

REPORTING PERIOD

Enter the appropriate calendar year, beginning January 1 and ending December 31.

FACILITY IDENTIFICATION

Enter the complete name of your facility (and company identifier where appropriate).

Enter the full street address or state road. If a street address is not available, enter other appropriate identifiers that describe the physical location of your facility (e.g., longitude and latitude). Include city, county, state, and zip code.

Enter the primary Standard Industrial Classification (SIC) code and the Dun & Bradstreet number for your facility. The financial officer of your facility should be able to provide the Dun & Bradstreet number. If your firm does not have this information, contact the State or regional office of Dun & Bradstreet to obtain your facility number or have one assigned.

OWNER/OPERATOR

Enter the owner's or operator's full name, mailing address, and phone number.

EMERGENCY CONTACT

Enter the name, title, and work phone number of at least one local person or office that can act as a referral if emergency responders need assistance in responding to a chemical accident at the facility.

Provide an emergency phone number where such emergency information will be available 24 hours a day, every day. This requirement is mandatory. The facility must make some arrangement to ensure that a 24 hour contact is available.

IDENTICAL INFORMATION

Check the box indicating identical information, located below the emergency contacts on the Tier One form, if the current information being reported is identical to that submitted last year. Chemical descriptions, amounts, and locations must be provided in this year's form, even if the information is identical to that submitted last year.

PHYSICAL AND HEALTH HAZARDS

Descriptions, Amounts, and Locations

This section requires aggregate information on chemicals by hazard categories as defined in 40 CFR 370.2. The two health hazard categories and three physical hazard categories are a consolidation of the 23 hazard categories defined in the OSHA Hazard Communication Standard, 29 CFR 1910.1200. For each hazard type, indicate the total amounts and general locations of all applicable chemicals present at your facility during the past year.

Hazard Category Comparison
For Reporting Under Sections 311 and 312

EPA's Hazard Categories	OSHA's Hazard Categories
Fire Hazard	Flammable Combustion Liquid Pyrophoric Oxidizer
Sudden Release of Pressure	Explosive Compressed Gas
Reactive	Unstable Reactive Organic Peroxide Water Reactive
Immediate (Acute) Health Hazards	Highly Toxic Toxic Irritant Sensitizer Corrosive Other hazardous chemicals with an adverse effect with short term exposure
Delayed (Chronic) Health Hazard	Carcinogens Other hazardous chemicals with an adverse effect with long term exposure

- What units should I use?

  Calculate all amounts as *weight in pounds*. To convert gas or liquid volume to weight in pounds, multiply by an appropriate density factor.

INSTRUCTIONS

Please read these instructions carefully. Print or type all responses.

- What about mixtures?

  If a chemical is part of a mixture, *you have the option* of reporting either the weight of the entire mixture or only the portion of the mixture that is a particular hazardous chemical (e.g., if a hazardous solution weighs 100 lbs. but is composed of only 5% of a particular hazardous chemical, you can indicate either 100 lbs. of the mixture or 5 lbs. of the hazardous chemical).

  The option used for each mixture must be consistent with the option used in your Section 311 reporting.

  Because EHSs are important to Section 303 planning, EHSs have lower thresholds. The amount of an EHS at a facility (both pure EHS substances and EHSs in mixtures) must be aggregated for purposes of threshold determination. It is suggested that the aggregation calculation be done as a first step in making the threshold determination. Once you determine whether a threshold has been reached for an EHS, you should report either the total weight of the EHS at your facility, or the weight of each mixture containing the EHS.

- Where do I count a chemical that is a fire and reactive physical hazard and an immediate (acute) health hazard?

  Add the chemical's weight to your totals for all three hazard categories and include its location in all three categories. Many chemicals fall into more than one hazard category.

MAXIMUM AMOUNT

The amounts of chemicals you have on hand may vary throughout the year. The peak weights -- greatest single-day weights during the year -- are added together in this column to determine the maximum weight for each hazard type. Since the peaks for different chemicals often occur on different days, this maximum amount will seem artificially high.

To complete this and the following sections, you may choose to use the Tier Two form as a worksheet.

To determine the Maximum Amount:

1. List all of your reportable hazardous chemicals individually.
2. For each chemical...
   a. Indicate all physical and health hazards that the chemical presents. Include all chemicals, even if they are present for only a short period of time during the year.
   b. Estimate the maximum weight in pounds that was present at your facility on any single day of the reporting period.
3. For each hazard type -- beginning with Fire and repeating for all physical and health hazard types...
   a. Add the maximum weights of all chemicals you indicated as the particular hazard type.
   b. Look at the Reporting Ranges at the bottom of the Tier One form. Find the appropriate range value code.
   c. Enter this range value as the Maximum Amount.

EXAMPLE:

You are using the Tier Two form as a worksheet and have listed raw weights in pounds for each of your hazardous chemicals. You have marked an X in the Immediate (acute) hazard column for phenol and sulfuric acid. The maximum amount raw weight you listed were 10,000 lbs. and 500 lbs. respectively. You add these together to reach a total of 10,500 lbs. Then you look at the Reporting Range at the bottom of your Tier One form and find that the value of 04 corresponds to 10,500 lbs. Enter 04 as your Maximum Amount for Immediate (acute) hazards materials.

You also marked an X in the Fire hazard box for phenol. When you calculate your Maximum Amount totals for fire hazards, add the 10,000 lb. weight again.

AVERAGE DAILY AMOUNT

This column should represent the average daily amount of chemicals *of each hazard type* that were present at or above applicable thresholds at your facility at any point during the year.

To determine this amount:

1. List all of your reportable hazardous chemicals individually (same as for Maximum Amount).
2. For each chemical...
   a. Indicate all physical and health hazards that the chemical presents (same as for Maximum Amount).
   b. Estimate the average weight in pounds that was present at your facility throughout the year. To do this, total all daily weights and divide by the number of days the chemical was present on the site.
3. For each hazard type -- beginning with Fire and repeating for all physical and health hazards...
   a. Add the average weights of all chemicals you indicated for the particular hazard type.
   b. Look at the Reporting Ranges at the bottom of the Tier One form. Find the appropriate range value code.
   c. Enter this range value as the Average Daily Amount.

Please read these instructions carefully. Print or type all responses.

EXAMPLE:

You are using the Tier Two form, and have marked an X in the Immediate (acute) hazard column for nicotine and phenol. Nicotine is present at your facility 100 days during the year, and the sum of the daily weights is 100,000 lbs. By dividing 100,000 lbs. by 100 days on-site, you calculate an Average Daily Amount of 1,000 lbs. for nicotine. Phenol is present at your facility 50 days during the year, and the sum of the daily weights is 10,000 lbs. By dividing 10,000 lbs. by 50 days on-site, you calculate an Average Daily Amount of 200 lbs. for phenol. You then add the two average daily amounts together to reach a total of 1,200 lbs. Then you look at the Reporting Range on your Tier One form and find that the value 03 corresponds to 1,200 lbs. Enter 03 as your Average Daily Amount for Immediate (acute) Hazard.

You also marked an X in the Fire hazard column for phenol. When you calculate your Average Daily Amount for fire hazards, use the 200 lb. weight again.

NUMBER OF DAYS ON-SITE

Enter the greatest number of days that a single chemical within that hazard category was present on-site.

EXAMPLE:

At your facility, nicotine is present for 100 days and phosgene is present for 150 days. Enter 150 in the space provided.

GENERAL LOCATION

Enter the general location within your facility where each hazard may be found. General locations should include the names or identifications of buildings, tank fields, lots, sheds, or other such areas.

For each hazard type, list the locations of all applicable chemicals. As an alternative you may also attach a site plan and list the site coordinates related to the appropriate locations. If you do so, check the Site Plan box.

EXAMPLE:

On your worksheet you have marked an X in the Fire hazard column for acetone and butane. You noted that these are kept in steel drums in Room C of the Main Building, and in pressurized cylinders in Storage Shed 13, respectively. You could enter Main Building and Storage Shed 13 as the General Locations of your fire hazards. However, you choose to attach a site plan and list coordinates. Check the Site Plan box at the top of the column and enter site coordinates for the Main Building and Storage Shed 13 under General Locations.

If you need more space to list locations, attach an additional Tier One form and continue your list on the proper line. Number all pages.

CERTIFICATION

Instructions for this section are included on page one of these instructions.

Sample Tier II Form

Revised November 1990

Form Approved OMB No. 2050–0072

Tier Two

EMERGENCY AND HAZARDOUS CHEMICAL INVENTORY

Specific Information by Chemical

Facility Identification

Name ______

Street ______

City ______ County ______ State ______ Zip ______

SIC Code ☐☐☐☐ Dun & Brad Number ☐☐-☐☐☐-☐☐☐☐

FOR OFFICIAL USE ONLY

ID # ______

Date Received ______

Owner/Operator Name

Name ______ Phone ( )

Mail Address ______

Emergency Contact

Name ______ Title ______

Phone ( ) 24 Hr. Phone ( )

Name ______ Title ______

Phone ( ) 24 Hr. Phone ( )

Important: Read all instructions before completing form

Reporting Period From January 1 to December 31, 19____

☐ Check if information below is identical to the information submitted last year.

Chemical Description	Physical and Health Hazards (check all that apply)	Inventory	Container Type / Pressure / Temperature	Storage Codes and Locations (Non-Confidential) — *Storage Locations*	Optional
CAS ☐☐☐☐☐☐-☐☐-☐ Trade Secret ☐ Chem. Name ______ *Check all that apply:* ☐ Pure ☐ Mix ☐ Solid ☐ Liquid ☐ Gas ☐ EHS EHS Name ______	☐ Fire ☐ Sudden Release of Pressure ☐ Reactivity ☐ Immediate (acute) ☐ Delayed (chronic)	☐☐ Max. Daily Amount (code) ☐☐ Avg. Daily Amount (code) ☐☐☐ No. of Days On-site (days)			☐
CAS ☐☐☐☐☐☐-☐☐-☐ Trade Secret ☐ Chem. Name ______ *Check all that apply:* ☐ Pure ☐ Mix ☐ Solid ☐ Liquid ☐ Gas ☐ EHS EHS Name ______	☐ Fire ☐ Sudden Release of Pressure ☐ Reactivity ☐ Immediate (acute) ☐ Delayed (chronic)	☐☐ Max. Daily Amount (code) ☐☐ Avg. Daily Amount (code) ☐☐☐ No. of Days On-site (days)			☐
CAS ☐☐☐☐☐☐-☐☐-☐ Trade Secret ☐ Chem. Name ______ *Check all that apply:* ☐ Pure ☐ Mix ☐ Solid ☐ Liquid ☐ Gas ☐ EHS EHS Name ______	☐ Fire ☐ Sudden Release of Pressure ☐ Reactivity ☐ Immediate (acute) ☐ Delayed (chronic)	☐☐ Max. Daily Amount (code) ☐☐ Avg. Daily Amount (code) ☐☐☐ No. of Days On-site (days)			☐

Certification *(Read and sign after completing all sections)*

I certify under penalty of law that I have personally examined and am familiar with the information submitted in pages one through ______, and that based on my inquiry of those individuals responsible for obtaining the information, I believe that the submitted information is true, accurate, and complete.

Name and official title of owner/operator OR owner/operator's authorized representative ______ Signature ______ Date signed ______

Optional Attachments

☐ I have attached a site plan

☐ I have attached a list of site coordinate abbreviations

☐ I have attached a description of dikes and other safeguard measures

Form Approved OMB No. 2050–0072

Tier Two
EMERGENCY AND HAZARDOUS CHEMICAL INVENTORY
Specific Information by Chemical

Facility Identification

Name ______
Street ______
City ______ County ______ State ______ Zip ______
SIC Code ☐☐☐☐ Dun & Brad Number ☐☐-☐☐☐-☐☐☐☐

FOR OFFICIAL USE ONLY
ID # ______
Date Received ______

Owner/Operator Name

Name ______ Phone ( )______
Mail Address ______

Emergency Contact

Name ______ Title ______
Phone ( )______ 24 Hr. Phone ( )______

Name ______ Title ______
Phone ( )______ 24 Hr. Phone ( )______

Important: Read all instructions before completing form

Reporting Period From January 1 to December 31, 19____

☐ Check if information below is identical to the information submitted last year.

Confidential Location Information Sheet

	Container Type	Pressure	Temperature	Storage Codes and Locations (Confidential) *Storage Locations*	Optional
CAS # ☐☐☐☐☐☐ ☐☐ ☐ Chem. Name					☐
CAS # ☐☐☐☐☐☐ ☐☐ ☐ Chem. Name					☐
CAS # ☐☐☐☐☐☐ ☐☐ ☐ Chem. Name					☐

Certification *(Read and sign after completing all sections)*

I certify under penalty of law that I have personally examined and am familiar with the information submitted in pages one through ______, and that based on my inquiry of those individuals responsible for obtaining the information, I believe that the submitted information is true, accurate, and complete.

Name and official title of owner/operator OR owner/operator's authorized representative ______ Signature ______ Date signed ______

Optional Attachments

☐ I have attached a site plan
☐ I have attached a list of site coordinate abbreviations
☐ I have attached a description of dikes and other safeguard measures

TIER TWO INSTRUCTIONS

GENERAL INFORMATION

Submission of this Tier Two form (when requested) is required by Title III of the Superfund Amendments and Reauthorization Act of 1986, Section 312, Public Law 99-499, codified at 42 U.S.C. Section 11022. The purpose of this Tier Two form is to provide State and local officials and the public with specific information on hazardous chemicals present at your facility during the past year.

CERTIFICATION

The owner or operator or the officially designated representative of the owner or operator must certify that all information included in the Tier Two submission is true, accurate, and complete. On the first page of the Tier Two report, enter your full name and official title. Sign your name and enter the current date. Also, enter the total number of pages included in the Confidential and Non-Confidential Information Sheets as well as all attachments. An original signature is required on at least the first page of the submission. Submissions to the SERC, LEPC, and fire department must each contain an original signature on at least the first page. Subsequent pages must contain either an original signature, a photocopy of the original signature, or a signature stamp. Each page must contain the date on which the original signature was affixed to the first page of the submission and the total number of pages in the submission.

YOU MUST PROVIDE ALL INFORMATION REQUESTED ON THIS FORM TO FULFILL TIER TWO REPORTING REQUIREMENTS.

This form may also be used as a worksheet for completing the Tier One form or may be submitted in place of the Tier One form.

WHO MUST SUBMIT THIS FORM

Section 312 of Title III requires that the owner or operator of a facility submit this Tier Two form if so requested by a State emergency response commission, a local emergency planning committee, or a fire department with jurisdiction over the facility.

This request may apply to the owner or operator of any facility that is required, under regulations implementing the Occupational Safety and Health Act of 1970, to prepare or have available a Material Safety Data Sheet (MSDS) for a hazardous chemical present at the facility. MSDS requirements are specified in the Occupational Safety and Health Administration (OSHA) Hazard Communication Standard, found in Title 29 of the Code of Federal Regulations at §1910.1200.

This form does not have to be submitted if all of the chemicals located at your facility are excluded under Section 311(e) of Title III.

WHAT CHEMICALS ARE INCLUDED

If you are submitting Tier Two forms in lieu of Tier One, you must report the required information on this Tier Two form for each hazardous chemical present at your facility in quantities equal to or greater than established threshold amounts (discussed below), unless the chemicals are excluded under Section 311(e) of Title III. Hazardous chemicals are any substance for which your facility must maintain an MSDS under OSHA's Hazard Communication Standard.

If you elect to submit Tier One rather than Tier Two, you may still be required to submit Tier Two information upon request.

WHAT CHEMICALS ARE EXCLUDED

Section 311(e) of Title III excludes the following substances:

(i) Any food, food additive, color additive, drug, or cosmetic regulated by the Food and Drug Administration;

(ii) Any substance present as a solid in any manufactured item to the extent exposure to the substance does not occur under normal conditions of use;

(iii) Any substance to the extent it is used for personal, family, or household purposes, or is present in the same form and concentration as a product packaged for distribution and use by the general public;

(iv) Any substance to the extent it is used in a research laboratory or a hospital or other medical facility under the direct supervision of a technically qualified individual;

(v) Any substance to the extent it is used in routine agricultural operations or is a fertilizer held for sale by a retailer to the ultimate customer.

OSHA regulations, Section 1910.1200(b), stipulate exemptions from the requirement to prepare or have available an MSDS.

REPORTING THRESHOLDS

Minimum thresholds have been established for Tier One/Tier Two reporting under Title III, Section 312. These thresholds are as follows:

For Extremely Hazardous Substances (EHSs) designated under section 302 of Title III, the reporting threshold is 500 pounds (or 227 kg.) or the threshold planning quantity (TPQ), whichever is lower;

For all other hazardous chemicals for which facilities are required to have or prepare an MSDS, the minimum reporting threshold is 10,000 pounds (or 4,540 kg.).

You need to report hazardous chemicals that were present at your facility at any time during the previous calendar year at levels that equal or exceed these thresholds. For instructions on threshold determinations for components of mixtures, see "What About Mixtures?" on page 2 of these instructions.

A requesting official may limit the responses required under Tier Two by specifying particular chemicals or groups of chemicals. Such requests apply to hazardous chemicals regardless of established thresholds.

INSTRUCTIONS

Please read these instructions carefully. Print or type all responses.

WHEN TO SUBMIT THIS FORM

Owners or operators of facilities that have hazardous chemicals on hand in quantities equal to or greater than set threshold levels must submit either Tier One or Tier Two forms by March 1.

If you choose to submit Tier One, rather than Tier Two, be aware that you may have to submit Tier Two information later, upon request of an authorized official. You must submit the Tier Two form within 30 days of receipt of a written request.

WHERE TO SUBMIT THIS FORM

Send either a completed Tier One form or Tier Two form(s) to each of the following organizations:

1. Your State Emergency Response Commission.
2. Your Local Emergency Planning Committee.
3. The fire department with jurisdiction over your facility.

If a Tier Two form is submitted in response to a request, send the completed form to the requesting agency.

PENALTIES

Any owner or operator who violates any Tier Two reporting requirements shall be liable to the United States for a civil penalty of up to $25,000 for each such violation. Each day a violation continues shall constitute a separate violation.

If your Tier Two responses require more than one page use additional forms and fill in the page number at the top of the form.

REPORTING PERIOD

Enter the appropriate calendar year, beginning January 1 and ending December 31.

FACILITY IDENTIFICATION

Enter the full name of your facility (and company identifier where appropriate).

Enter the full street address or state road. If a street address is not available, enter other appropriate identifiers that describe the physical location of your facility (e.g., longitude and latitude). Include city, county, state, and zip code.

Enter the primary Standard Industrial Classification (SIC) code and the Dun & Bradstreet number for your facility. The financial officer of your facility should be able to provide the Dun & Bradstreet number. If your firm does not have this information, contact the State or regional office of Dun & Bradstreet to obtain your facility number or have one assigned.

OWNER/OPERATOR

Enter the owner's or operator's full name, mailing address, and phone number.

EMERGENCY CONTACT

Enter the name, title, and work phone number of at least one local person or office who can act as a referral if emergency responders need assistance in responding to a chemical accident at the facility.

Provide an emergency phone number where such emergency information will be available 24 hours a day, every day. This requirement is mandatory. The facility must make some arrangement to ensure that a 24 hour contact is available.

IDENTICAL INFORMATION

Check the box indicating identical information, located below the emergency contacts on the Tier Two form, if the current chemical information being reported is identical to that submitted last year. Chemical descriptions, hazards, amounts, and locations must be provided in this year's form, even if the information is identical to that submitted last year.

CHEMICAL INFORMATION: Description, Hazards, Amounts, and Locations

The main section of the Tier Two form requires specific information on amounts and locations of hazardous chemicals, as defined in the OSHA Hazard Communication Standard.

If you choose to indicate that all of the information on a specific hazardous chemical is identical to that submitted last year, check the appropriate optional box provided at the right side of the storage codes and locations on the Tier Two form. Chemical descriptions, hazards, amounts, and locations must be provided even if the information is identical to that submitted last year.

- What units should I use?

  Calculate all amounts as *weight in pounds.* To convert gas or liquid volume to weight in pounds, multiply by an appropriate density factor.

- What about mixtures?

  If a chemical is part of a mixture, *you have the option* of reporting either the weight of the entire mixture or only the portion of the mixture that is a particular hazardous chemical (e.g., if a hazardous solution weighs 100 lbs. but is composed of only 5% of a particular hazardous chemical, you can indicate either 100 lbs. of the mixture *or* 5 lbs. of the chemical).

  The option used for each mixture must be consistent with the option used in your Section 311 reporting.

  Because EHSs are important to Section 303 planning, EHSs have lower thresholds. The amount of an EHS at a facility (both pure EHS substances and EHSs in mixtures) must be aggregated for purposes of threshold determination. It is suggested that the aggregation calculation be done as a first step in making the threshold determination. Once you determine whether a threshold for an EHS has been reached, you should report either the total weight of the EHS at your facility, or the weight of each mixture containing the EHS.

CHEMICAL DESCRIPTION

1. Enter the Chemical Abstract Service registry number (CAS). For mixtures, enter the CAS number of the mixture as a whole if it has been assigned a number distinct from its constituents. For a mixture that has no CAS number, leave this item blank or report the CAS numbers of as many constituent chemicals as possible.

If you are withholding the name of a chemical in accordance with criteria specified in Title III, Section 322, enter the generic class or category that is structurally descriptive of the chemical (e.g., list toulene diisocyanate as organic isocyanate) and check the box marked Trade Secret. Trade secret information should be submitted to EPA and must include a substantiation. Please refer to EPA's final regulation on trade secrecy (53 FR 28772, July 29, 1988) for detailed information on how to submit trade secrecy claims.

2. Enter the chemical name or common name of each hazardous chemical.

3. Check box for *ALL* applicable descriptors: pure or mixture; *and* solid, liquid, or gas; and whether the chemical is or contains an EHS.

4. If the chemical is a mixture containing an EHS, enter the chemical name of each EHS in the mixture.

EXAMPLE:

You have pure chlorine gas on hand, as well as two mixtures that contain liquid chlorine. You write "chlorine" and enter the CAS number. Then you check "pure" *and* "mix" -- as well as "liquid" *and* "gas".

PHYSICAL AND HEALTH HAZARDS

For each chemical you have listed, check all the physical and health hazard boxes that apply. These hazard categories are defined in 40 CFR 370.2. The two health hazard categories and three physical hazard categories are a consolidation of the 23 hazard categories defined in the OSHA Hazard Communication Standard, 29 CFR 1910.1200.

Hazard Category Comparison For Reporting Under Sections 311 and 312

EPA's Hazard Categories	OSHA's Hazard Categories
Fire Hazard	Flammable Combustion Liquid Pyrophoric Oxidizer
Sudden Release of Pressure	Explosive Compressed Gas
Reactive	Unstable Reactive Organic Peroxide Water Reactive
Immediate (Acute) Health Hazards	Highly Toxic Toxic Irritant Sensitizer Corrosive Other hazardous chemicals with an adverse effect with short term exposure
Delayed (Chronic) Health Hazard	Carcinogens Other hazardous chemicals with an adverse effect with long term exposure

MAXIMUM AMOUNT

1. For each hazardous chemical, estimate the greatest amount present at your facility on any single day during the reporting period.

2. Find the appropriate range value code in Table I.

3. Enter this range value as the Maximum Amount.

Table I REPORTING RANGES

Range Value	Weight Range in Pounds From...	To...
01	0	99
02	100	999
03	1,000	9,999
04	10,000	99,999
05	100,000	999,999
06	1,000,000	9,999,999
07	10,000,000	49,999,999
08	50,000,000	99,999,999
09	100,000,000	499,999,999
10	500,000,000	999,999,999
11	1 billion	higher than 1 billion

If you are using this form as a worksheet for completing Tier One, enter the actual weight in pounds in the shaded space below the response blocks. Do this for both Maximum Amount and Average Daily Amount.

> **EXAMPLE:**
>
> You received one large shipment of a solvent mixture last year. The shipment filled five 5,000 - gallon storage tanks. You know that the solvent contains 10% benzene, which is a hazardous chemical.
>
> You figure that 10% of 25,000 gallons is 2,500 gallons. You also know that the density of benzene is 7.29 pounds per gallon, so you multiply 2,500 gallons by 7.29 pounds per gallon to get a weight of 18,225 pounds.
>
> Then you look at Table I and find that the range value 04 corresponds to 18,225. You enter 04 as the Maximum Amount.
>
> (If you are using the form as a worksheet for completing a Tier One form, you should write 18,255 in the shaded area.)

AVERAGE DAILY AMOUNT

1. For each hazardous chemical, estimate the average weight in pounds that was present at your facility during the year.

   To do this, total all daily weights and divide by the number of days the chemical was present on the site.
2. Find the appropriate range value in Table I.
3. Enter this range value as the Average Daily Amount.

> **EXAMPLE:**
>
> The 25,000-gallon shipment of solvent you received last year was gradually used up and completely gone in 315 days. The sum of the daily volume levels in the tank is 4,536,000 gallons. By dividing 4,536,000 gallons by 315 days on-site, you calculate an average daily amount of 14,400 gallons.
>
> You already know that the solvent contains 10% benzene, which is a hazardous chemical. Since 10% of 14,400 is 1,440, you figure that you had an average of 1,440 gallons of benzene. You also know that the density of benzene is 7.29 pounds per gallon, so you multiply 1,440 by 7.29 to get a weight of 10,500 pounds.
>
> Then you look at Table I and find that the range value 04 corresponds to 10,500. You enter 04 as the Average Daily Amount.
>
> (If you are using the form as a worksheet for completing a Tier One form, you should write 10,500 in the shaded area.)

NUMBER OF DAYS ON-SITE

Enter the number of days that the hazardous chemical was found on-site.

> **EXAMPLE:**
>
> The solvent composed of 10% benzene was present for 315 days at your facility. Enter 315 in the space provided.

STORAGE CODES AND STORAGE LOCATIONS

List all non-confidential chemical locations in this column, along with storage types/conditions associated with each location. Please note that a particluar chemical may be located in several places around the facility. Each row of boxes followed by a line represents a unique location for the same chemical.

Storage Codes: Indicate the types and conditions of storage present.

a. *Look at Table II.* For each location, find the appropriate storage type and enter the corresponding code in the first box.

b. *Look at Table III.* For each location, find the appropriate storage types for pressure and temperature conditions. Enter the applicable pressure code in the second box. Enter the applicable temperature code in the third box.

Table II – STORAGE TYPES

CODES	Types of Storage
A	Above ground tank
B	Below ground tank
C	Tank inside building
D	Steel drum
E	Plastic or non-metallic drum
F	Can
G	Carboy
H	Silo
I	Fiber drum
J	Bag
K	Box
L	Cylinder
M	Glass bottles or jugs
N	Plastic bottles or jugs
O	Tote bin
P	Tank wagon
Q	Rail car
R	Other

Table III – PRESSURE AND TEMPERATURE CONDITIONS

CODES	Storage Conditions
	(PRESSURE)
1	Ambient pressure
2	Greater than ambient pressure
3	Less than ambient pressure
	(TEMPERATURE)
4	Ambient temperature
5	Greater than ambient temperature
6	Less than ambient temperature but not cryogenic
7	Cryogenic conditions

> **EXAMPLE:**
>
> The benzene in the main building is kept in a tank inside the building, at ambient pressure and less than ambient temperature.
>
> Table II shows you that the code for a tank inside a building is C. Table III shows you that the code for ambient pressure is 1, and the code for less than ambient temperature is 6.
>
> You enter: | C | 1 | 6 |

Storage Locations:

Provide a brief description of the precise location of the chemical, so that emergency responders can locate the area easily. You may find it advantageous to provide the optional site plan or site coordinates as explained below.

For each chemical, indicate at a minimum the building or lot. Additionally, where practical, the room or area may be indicated. You may respond in narrative form with appropriate site coordinates or abbreviations.

If the chemical is present in more than one building, lot, or area location, continue your responses down the page as needed. If the chemical exists everywhere at the plant site simultaneously, you may report that the chemical is ubiquitous at the site.

Optional attachments: If you choose to attach one of the following, check the appropriate Attachments box at the bottom of the Tier Two form.

a. *A site plan* with site coordinates indicated for buildings, lots, areas, etc. throughout your facility.

b. *A list of site coordinate abbreviations* that correspond to buildings, lots, areas, etc. throughout your facility.

c. *A description of dikes and other safeguard measures* for storage locations throughout your facility.

EXAMPLE:

You have benzene in the main room of the main building, and in tank 2 in tank field 10. You attach a site plan with coordinates as follows: main building = G-2, tank field 10 = B-6. Fill in the Storage Location as follows:

B-6 [Tank 2] G-2 [Main Room]

CONFIDENTIAL INFORMATION

Under Title III, Section 324, you may elect to withhold location information on a specific chemical from disclosure to the public. If you choose to do so:

- Enter the word "confidential" in the Non-Confidential Location section of the Tier Two form on the first line of the storage locations..
- On a separate Tier Two Confidential Location Information Sheet, enter the name and CAS number of each chemical for which you are keeping the location confidential.
- Enter the appropriate location and storage information, as described above for non-confidential locations.
- Attach the Tier Two Confidential Location Information Sheet to the Tier Two form. This separates confidential locations from other information that will be disclosed to the public.

CERTIFICATION

Instructions for this section are included on page one of these instructions.

Section 313 Toxic Chemical List

SECTION 313 TOXIC CHEMICAL LIST FOR REPORTING YEAR 1991 (including Toxic Chemical Categories)

Specific toxic chemicals with CAS Number are listed in alphabetical order on this page. A list of the same chemicals in CAS Number order begins at the end of the alphabetical list of toxic chemicals. Covered toxic chemical categories follow.

Certain toxic chemicals listed in Table II have parenthetic "qualifiers." These qualifiers indicate that these toxic chemicals are subject to the section 313 reporting requirements if manufactured, processed, or otherwise used in a specific form. The following chemicals are reportable only if they are manufactured, processed, or otherwise used in the specific form(s) listed below:

Chemical	CAS Number	Qualifier
Aluminum (fume or dust)	7429-90-5	Only if it is in a fume or dust form.
Aluminum oxide (fibrous forms)	1344-28-1	Only if it is a fibrous form.
Ammonium nitrate (solution)	6484-52-2	Only if it is in a solution.
Ammonium sulfate (solution)	7783-20-2	Only if it is in a solution.
Asbestos (friable)	1332-21-4	Only if it is a friable form.
Isopropyl alcohol (manufacturing - strong acid process, no supplier notification)	67-63-0	Only if it is being manufactured by the strong acid process.
Phosphorus (yellow or white)	7723-14-0	Only if it is a yellow or white form.
Saccharin (manufacturing, no supplier notification).	81-07-2	Only if it is being manufactured.
Vanadium (fume or dust)	7440-62-2	Only if it is in a fume or dust form.
Zinc (fume or dust)	7440-66-6	Only if it is in a fume or dust form.

[Note: Chemicals may be added to or deleted from the list. The Emergency Planning and Community Right-to-Know Information Hotline, (800) 535-0202 or (703) 920-9877, will provide up-to-date information on the status of these changes. See Section B.4.b of the instructions for more information on the de minimis values listed below.]

*C.I. means "Color Index"

a. Alphabetical Chemical List

CAS Number	Toxic Chemical Name	De Minimis Concentration
75-07-0	Acetaldehyde	0.1
60-35-5	Acetamide	0.1
67-64-1	Acetone	1.0
75-05-8	Acetonitrile	1.0
53-96-3	2-Acetylaminofluorene	0.1
107-02-8	Acrolein	1.0
79-06-1	Acrylamide	0.1
79-10-7	Acrylic acid	1.0
107-13-1	Acrylonitrile	0.1
309-00-2	Aldrin {1,4:5,8-Dimethanonaphthalene, 1,2,3,4,10,10-hexachloro-1,4,4a, 5,8,8a-hexahydro-(1.alpha., 4.alpha.,4a.beta.,5.alpha., 8.alpha.,8a.beta.)-}	1.0
107-18-6	Allyl alcohol	1.0
107-05-1	Allyl chloride	1.0
7429-90-5	Aluminum (fume or dust)	1.0
1344-28-1	Aluminum oxide (fibrous forms)	0.1
117-79-3	2-Aminoanthraquinone	0.1
60-09-3	4-Aminoazobenzene	0.1
92-67-1	4-Aminobiphenyl	0.1
82-28-0	1-Amino-2-methylanthraquinone	0.1
7664-41-7	Ammonia	1.0
6484-52-2	Ammonium nitrate (solution)	1.0
7783-20-2	Ammonium sulfate (solution)	1.0
62-53-3	Aniline	1.0
90-04-0	o-Anisidine	0.1
104-94-9	p-Anisidine	1.0
134-29-2	o-Anisidine hydrochloride	0.1
120-12-7	Anthracene	1.0
7440-36-0	Antimony	1.0
7440-38-2	Arsenic	0.1
1332-21-4	Asbestos (friable)	0.1
7440-39-3	Barium	1.0
98-87-3	Benzal chloride	1.0
55-21-0	Benzamide	1.0
71-43-2	Benzene	0.1
92-87-5	Benzidine	0.1
98-07-7	Benzoic trichloride {Benzotrichloride}	0.1
98-88-4	Benzoyl chloride	1.0
94-36-0	Benzoyl peroxide	1.0
100-44-7	Benzyl chloride	1.0
7440-41-7	Beryllium	0.1

CAS Number	Toxic Chemical Name	De Minimis Concentration
92-52-4	Biphenyl	1.0
111-44-4	Bis(2-chloroethyl) ether	1.0
542-88-1	Bis(chloromethyl) ether	
0.1108-60-1	Bis(2-chloro-1-methylethyl) ether	1.0
103-23-1	Bis(2-ethylhexyl) adipate	1.0
353-59-3	Bromochlorodifluoromethane {Halon 1211}	1.0
75-25-2	Bromoform {Tribromomethane}	1.0
74-83-9	Bromomethane {Methyl bromide}	1.0
75-63-8	Bromotrifluoromethane {Halon 1301}	1.0
106-99-0	1,3-Butadiene	0.1
141-32-2	Butyl acrylate	1.0
71-36-3	n-Butyl alcohol	1.0
78-92-2	sec-Butyl alcohol	1.0
75-65-0	tert-Butyl alcohol	1.0
85-68-7	Butyl benzyl phthalate	1.0
106-88-7	1,2-Butylene oxide	1.0
123-72-8	Butyraldehyde	1.0
4680-78-8	C.I. Acid Green 3*	1.0
569-64-2	C.I. Basic Green 4*	1.0
989-38-8	C.I. Basic Red 1*	0.1
1937-37-7	C.I. Direct Black 38*	0.1
2602-46-2	C.I. Direct Blue 6*	0.1
16071-86-6	C.I. Direct Brown 95*	0.1
2832-40-8	C.I. Disperse Yellow 3*	1.0
3761-53-3	C.I. Food Red 5*	0.1
81-88-9	C.I. Food Red 15*	0.1
3118-97-6	C.I. Solvent Orange 7*	1.0
97-56-3	C.I. Solvent Yellow 3*	0.1
842-07-9	C.I. Solvent Yellow 14*	0.1
492-80-8	C.I. Solvent Yellow 34* {Aurimine}	0.1
128-66-5	C.I. Vat Yellow 4*	1.0
7440-43-9	Cadmium	0.1
156-62-7	Calcium cyanamide	1.0
133-06-2	Captan {1H-Isoindole-1,3(2H)-dione, 3a,4,7,7a-tetrahydro-2-[(trichloromethyl)thio]-}	1.0
63-25-2	Carbaryl {1-Naphthalenol, methylcarbamate}	1.0
75-15-0	Carbon disulfide	1.0
56-23-5	Carbon tetrachloride	0.1
463-58-1	Carbonyl sulfide	1.0

*C.I. means "Color Index"

CAS Number	Toxic Chemical Name	De Minimis Concentration
120-80-9	Catechol	1.0
133-90-4	Chloramben {Benzoic acid, 3-amino-2,5-dichloro-}	1.0
57-74-9	Chlordane {4,7-Methanoindan, 1,2,4,5,6,7,8,8-octachloro-2,3,3a,4,7,7a-hexahydro-}	1.0
7782-50-5	Chlorine	1.0
10049-04-4	Chlorine dioxide	1.0
79-11-8	Chloroacetic acid	1.0
532-27-4	2-Chloroacetophenone	1.0
108-90-7	Chlorobenzene	1.0
510-15-6	Chlorobenzilate {Benzeneacetic acid,4-chloro-.alpha.-(4-chlorophenyl)-.alpha.-hydroxy-,ethyl ester}	1.0
75-00-3	Chloroethane {Ethyl chloride}	1.0
67-66-3	Chloroform	0.1
74-87-3	Chloromethane {Methyl chloride}	1.0
107-30-2	Chloromethyl methyl ether	0.1
126-99-8	Chloroprene	1.0
1897-45-6	Chlorothalonil {1,3-Benzenedicarbonitrile, 2,4,5,6-tetrachloro-}	1.0
7440-47-3	Chromium	0.1
7440-48-4	Cobalt	1.0
7440-50-8	Copper	1.0
8001-58-9	Creosote	0.1
120-71-8	p-Cresidine	0.1
1319-77-3	Cresol (mixed isomers)	1.0
108-39-4	m-Cresol	1.0
95-48-7	o-Cresol	1.0
106-44-5	p-Cresol	1.0
98-82-8	Cumene	1.0
80-15-9	Cumene hydroperoxide	1.0
135-20-6	Cupferron {Benzeneamine, N-hydroxy-N-nitroso, ammonium salt}	0.1
110-82-7	Cyclohexane	1.0
94-75-7	2,4-D {Acetic acid, (2,4-dichlorophenoxy)-}	1.0
1163-19-5	Decabromodiphenyl oxide	1.0
2303-16-4	Diallate {Carbamothioic acid, bis(1-methylethyl)-, S-(2,3-dichloro-2-propenyl) ester}	1.0

CAS Number	Toxic Chemical Name	De Minimis Concentration
615-05-4	2,4-Diaminoanisole	0.1
39156-41-7	2,4-Diaminoanisole sulfate	0.1
101-80-4	4,4'-Diaminodiphenyl ether	0.1
25376-45-8	Diaminotoluene (mixed isomers)	0.1
95-80-7	2,4-Diaminotoluene	0.1
334-88-3	Diazomethane	1.0
132-64-9	Dibenzofuran	1.0
96-12-8	1,2-Dibromo-3-chloropropane {DBCP}	0.1
106-93-4	1,2-Dibromoethane {Ethylene dibromide}	0.1
124-73-2	Dibromotetrafluoroethane {Halon 2402}	1.0
84-74-2	Dibutyl phthalate	1.0
25321-22-6	Dichlorobenzene (mixed isomers)	0.1
95-50-1	1,2-Dichlorobenzene	1.0
541-73-1	1,3-Dichlorobenzene	1.0
106-46-7	1,4-Dichlorobenzene	0.1
91-94-1	3,3'-Dichlorobenzidine	0.1
75-27-4	Dichlorobromomethane	1.0
75-71-8	Dichlorodifluoromethane (CFC-12)	1.0
107-06-2	1,2-Dichloroethane {Ethylene dichloride}	0.1
540-59-0	1,2-Dichloroethylene	1.0
75-09-2	Dichloromethane {Methylene chloride}	0.1
120-83-2	2,4-Dichlorophenol	1.0
78-87-5	1,2-Dichloropropane	1.0
78-88-6	2,3-Dichloropropene	1.0
542-75-6	1,3-Dichloropropylene	0.1
76-14-2	Dichlorotetrafluoroethane (CFC-114)	1.0
62-73-7	Dichlorvos {Phosphoric acid, 2,2-dichloroethenyl dimethyl ester}	1.0
115-32-2	Dicofol {Benzenemethanol, 4-chloro-.alpha.-(4-chlorophenyl)-.alpha.- (trichloromethyl)-}	1.0
1464-53-5	Diepoxybutane	0.1
111-42-2	Diethanolamine	1.0
117-81-7	Di-(2-ethylhexyl) phthalate {DEHP}	0.1
84-66-2	Diethyl phthalate	1.0
64-67-5	Diethyl sulfate	0.1
119-90-4	3,3'-Dimethoxybenzidine	0.1
60-11-7	4-Dimethylaminoazobenzene	0.1

*C.I. means "Color Index"

CAS Number	Toxic Chemical Name	De Minimis Concentration
119-93-7	3,3'-Dimethylbenzidine {o-Tolidine}	0.1
79-44-7	Dimethylcarbamyl chloride	0.1
57-14-7	1,1-Dimethyl hydrazine	0.1
105-67-9	2,4-Dimethylphenol	1.0
131-11-3	Dimethyl phthalate	1.0
77-78-1	Dimethyl sulfate	0.1
99-65-0	m-Dinitrobenzene	1.0
528-29-0	o-Dinitrobenzene	1.0
100-25-4	p-Dinitrobenzene	1.0
534-52-1	4,6-Dinitro-o-cresol	1.0
51-28-5	2,4-Dinitrophenol	1.0
121-14-2	2,4-Dinitrotoluene	1.0
606-20-2	2,6-Dinitrotoluene	1.0
25321-14-6	Dinitrotoluene (mixed isomers)	1.0
117-84-0	n-Dioctyl phthalate	1.0
123-91-1	1,4-Dioxane	0.1
122-66-7	1,2-Diphenylhydrazine {Hydrazobenzene}	0.1
106-89-8	Epichlorohydrin	0.1
110-80-5	2-Ethoxyethanol	1.0
140-88-5	Ethyl acrylate	0.1
100-41-4	Ethylbenzene	1.0
541-41-3	Ethyl chloroformate	1.0
74-85-1	Ethylene	1.0
107-21-1	Ethylene glycol	1.0
151-56-4	Ethyleneimine {Aziridine}	0.1
75-21-8	Ethylene oxide	0.1
96-45-7	Ethylene thiourea	0.1
2164-17-2	Fluometuron {Urea, N,N-dimethyl-N'-[3-(trifluoromethyl)phenyl]-}	1.0
50-00-0	Formaldehyde	0.1
76-13-1	Freon 113 {Ethane, 1,1,2-trichloro-1,2,2-trifluoro-}	1.0
76-44-8	Heptachlor {1,4,5,6,7,8,8-Heptachloro-3a,4,7,7a-tetrahydro-4,7-methano-1H-indene}	1.0
118-74-1	Hexachlorobenzene	0.1
87-68-3	Hexachloro-1,3-butadiene	1.0
77-47-4	Hexachlorocyclopentadiene	1.0
67-72-1	Hexachloroethane	1.0
1335-87-1	Hexachloronaphthalene	1.0
680-31-9	Hexamethylphosphoramide	0.1
302-01-2	Hydrazine	0.1
10034-93-2	Hydrazine sulfate	0.1

CAS Number	Toxic Chemical Name	De Minimis Concentration
7647-01-0	Hydrochloric acid	1.0
74-90-8	Hydrogen cyanide	1.0
7664-39-3	Hydrogen fluoride	1.0
123-31-9	Hydroquinone	1.0
78-84-2	Isobutyraldehyde	1.0
67-63-0	Isopropyl alcohol (manufacturing-strong acid process, no supplier notification)	0.1
80-05-7	4,4'-Isopropylidenediphenol	1.0
120-58-1	Isosafrole	1.0
7439-92-1	Lead	0.1
58-89-9	Lindane {Cyclohexane,1,2,3,4,5,6-hexachloro-,(1.alpha.,2.alpha.,3.beta.,4.alpha.,5.alpha.,6.beta.)-}	0.1
108-31-6	Maleic anhydride	1.0
12427-38-2	Maneb {Carbamodithioic acid, 1,2-ethanediylbis-,manganese complex}	1.0
7439-96-5	Manganese	1.0
7439-97-6	Mercury	1.0
67-56-1	Methanol	1.0
72-43-5	Methoxychlor {Benzene, 1,1'-(2,2,2-trichloroethylidene)bis[4-methoxy-]}	1.0
109-86-4	2-Methoxyethanol	1.0
96-33-3	Methyl acrylate	1.0
1634-04-4	Methyl tert-butyl ether	1.0
101-14-4	4,4'-Methylenebis (2-chloroaniline) {MBOCA}	0.1
101-61-1	4,4'-Methylenebis (N,N-dimethyl) benzenamine	0.1
101-68-8	Methylenebis (phenylisocyanate) {MBI}	1.0
74-95-3	Methylene bromide	1.0
101-77-9	4,4'-Methylenedianiline	0.1
78-93-3	Methyl ethyl ketone	1.0
60-34-4	Methyl hydrazine	1.0
74-88-4	Methyl iodide	0.1
108-10-1	Methyl isobutyl ketone	1.0
624-83-9	Methyl isocyanate	1.0
80-62-6	Methyl methacrylate	1.0
90-94-8	Michler's ketone	0.1
1313-27-5	Molybdenum trioxide	1.0
76-15-3	(Mono)chloropentafluoroethane {CFC-115}	1.0

*C.I. means "Color Index"

CAS Number	Toxic Chemical Name	De Minimis Concentration
505-60-2	Mustard gas {Ethane, 1,1'-thiobis[2-chloro-]}	0.1
91-20-3	Naphthalene	1.0
134-32-7	alpha-Naphthylamine	0.1
91-59-8	beta-Naphthylamine	0.1
7440-02-0	Nickel	0.1
7697-37-2	Nitric acid	1.0
139-13-9	Nitrilotriacetic acid	0.1
99-59-2	5-Nitro-o-anisidine	0.1
98-95-3	Nitrobenzene	1.0
92-93-3	4-Nitrobiphenyl	0.1
1836-75-5	Nitrofen {Benzene, 2,4-dichloro-1-(4-nitrophenoxy)-}	0.1
51-75-2	Nitrogen mustard {2-Chloro-N-(2-chloroethyl)-N-methylethanamine}	0.1
55-63-0	Nitroglycerin	1.0
88-75-5	2-Nitrophenol	1.0
100-02-7	4-Nitrophenol	1.0
79-46-9	2-Nitropropane	0.1
156-10-5	p-Nitrosodiphenylamine	0.1
121-69-7	N,N-Dimethylaniline	1.0
924-16-3	N-Nitrosodi-n-butylamine	0.1
55-18-5	N-Nitrosodiethylamine	0.1
62-75-9	N-Nitrosodimethylamine	0.1
86-30-6	N-Nitrosodiphenylamine	1.0
621-64-7	N-Nitrosodi-n-propylamine	0.1
4549-40-0	N-Nitrosomethylvinylamine	0.1
59-89-2	N-Nitrosomorpholine	0.1
759-73-9	N-Nitroso-N-ethylurea	0.1
684-93-5	N-Nitroso-N-methylurea	0.1
16543-55-8	N-Nitrosonornicotine	0.1
100-75-4	N-Nitrosopiperidine	0.1
2234-13-1	Octachloronaphthalene	1.0
20816-12-0	Osmium tetroxide	1.0
56-38-2	Parathion {Phosphorothioic acid, O, O-diethyl-O-(4-nitrophenyl) ester}	1.0
87-86-5	Pentachlorophenol {PCP}	1.0
79-21-0	Peracetic acid	1.0
108-95-2	Phenol	1.0
106-50-3	p-Phenylenediamine	1.0
90-43-7	2-Phenylphenol	1.0
75-44-5	Phosgene	1.0
7664-38-2	Phosphoric acid	1.0
7723-14-0	Phosphorus (yellow or white)	1.0
85-44-9	Phthalic anhydride	1.0
88-89-1	Picric acid	1.0

CAS Number	Toxic Chemical Name	De Minimis Concentration
1336-36-3	Polychlorinated biphenyls {PCBs}	0.1
1120-71-4	Propane sultone	0.1
57-57-8	beta-Propiolactone	0.1
123-38-6	Propionaldehyde	1.0
114-26-1	Propoxur {Phenol, 2-(1-methylethoxy)-, methylcarbamate}	1.0
115-07-1	Propylene {Propene}	1.0
75-55-8	Propyleneimine	0.1
75-56-9	Propylene oxide	0.1
110-86-1	Pyridine	1.0
91-22-5	Quinoline	1.0
106-51-4	Quinone	1.0
82-68-8	Quintozene {Pentachloronitrobenzene}	1.0
81-07-2	Saccharin (manufacturing, no supplier notification) {1,2-Benzisothiazol-3(2H)-one, 1,1-dioxide}	0.1
94-59-7	Safrole	0.1
7782-49-2	Selenium	1.0
7440-22-4	Silver	1.0
100-42-5	Styrene	0.1
96-09-3	Styrene oxide	0.1
7664-93-9	Sulfuric acid	1.0
79-34-5	1,1,2,2-Tetrachloroethane	0.1
127-18-4	Tetrachloroethylene {Perchloroethylene}	0.1
961-11-5	Tetrachlorvinphos {Phosphoric acid, 2-chloro-1-(2,4,5-trichlorophenyl) ethenyl dimethyl ester}	1.0
7440-28-0	Thallium	1.0
62-55-5	Thioacetamide	0.1
139-65-1	4,4'-Thiodianiline	0.1
62-56-6	Thiourea	0.1
1314-20-1	Thorium dioxide	1.0
7550-45-0	Titanium tetrachloride	1.0
108-88-3	Toluene	1.0
584-84-9	Toluene-2,4-diisocyanate	0.1
91-08-7	Toluene-2,6-diisocyanate	0.1
26471-62-5	Toluenediisocyanate (mixed isomers)	0.1
95-53-4	o-Toluidine	0.1
636-21-5	o-Toluidine hydrochloride	0.1
8001-35-2	Toxaphene	0.1

*C.I. means "Color Index"

CAS Number	Toxic Chemical Name	De Minimis Concentration
68-76-8	Triaziquone {2,5-Cyclohexadiene-1,4-dione, 2,3,5-tris(1-aziridinyl)-}	0.1
52-68-6	Trichlorfon {Phosphonic acid,(2,2,2-trichloro-1-hydroxyethyl)-,dimethyl ester}	1.0
120-82-1	1,2,4-Trichlorobenzene	1.0
71-55-6	1,1,1-Trichloroethane {Methyl chloroform}	1.0
79-00-5	1,1,2-Trichloroethane	1.0
79-01-6	Trichloroethylene	1.0
75-69-4	Trichlorofluoromethane {CFC-11}	1.0
95-95-4	2,4,5-Trichlorophenol	1.0
88-06-2	2,4,6-Trichlorophenol	0.1
1582-09-8	Trifluralin {Benzenamine, 2,6-dinitro-N,N-dipropyl-4-(trifluoromethyl)-1}	1.0
95-63-6	1,2,4-Trimethylbenzene	1.0
126-72-7	Tris (2,3-dibromopropyl) phosphate	0.1
51-79-6	Urethane {Ethyl carbamate}	0.1
7440-62-2	Vanadium (fume or dust)	1.0
108-05-4	Vinyl acetate	1.0
593-60-2	Vinyl bromide	0.1
75-01-4	Vinyl chloride	0.1
75-35-4	Vinylidene chloride	1.0
1330-20-7	Xylene (mixed isomers)	1.0
108-38-3	m-Xylene	1.0
95-47-6	o-Xylene	1.0
106-42-3	p-Xylene	1.0
87-62-7	2,6-Xylidine	1.0
7440-66-6	Zinc (fume or dust)	1.0
12122-67-7	Zineb {Carbamodithioic acid, 1,2-ethanediylbis-, zinc complex}	1.0

*C.I. means "Color Index"

Standard Industrial Classification Codes

SIC CODES 20-39

20 Food and Kindred Products

2011 Meat packing plants
2013 Sausages and other prepared meat products
2015 Poultry slaughtering and processing
2021 Creamery butter
2022 Natural, processed, and imitation cheese
2023 Dry, condensed, and evaporated dairy products
2024 Ice cream and frozen desserts
2026 Fluid milk
2032 Canned specialties
2033 Canned fruits, vegetables, preserves, jams, and jellies
2034 Dried and dehydrated fruits, vegetables, and soup mixes
2035 Pickled fruits and vegetables, vegetable sauces and seasonings, and salad dressings
2037 Frozen fruits, fruit juices, and vegetables
2038 Frozen specialties, n.e.c.*
2041 Flour and other grain mill products
2043 Cereal breakfast foods
2044 Rice milling
2045 Prepared flour mixes and doughs
2046 Wet corn milling
2047 Dog and cat food
2048 Prepared feeds and feed ingredients for animals and fowls, except dogs and cats
2051 Bread and other bakery products, except cookies and crackers
2052 Cookies and crackers
2053 Frozen bakery products, except bread
2061 Cane sugar, except refining
2062 Cane sugar refining
2063 Beet sugar
2064 Candy and other confectionery products
2066 Chocolate and cocoa products
2067 Chewing gum
2068 Salted and roasted nuts and seeds
2074 Cottonseed oil mills
2075 Soybean oil mills
2076 Vegetable oil mills, n.e.c.*
2077 Animal and marine fats and oils
2079 Shortening, table oils, margarine, and other edible fats and oils, n.e.c.*
2082 Malt beverages
2083 Malt
2084 Wines, brandy, and brandy spirits
2085 Distilled and blended liquors
2086 Bottled and canned soft drinks and carbonated waters
2087 Flavoring extracts and flavoring syrups, n.e.c.*
2091 Canned and cured fish and seafoods
2092 Prepared fresh or frozen fish and seafoods
2095 Roasted coffee
2096 Potato chips, corn chips, and similar snacks
2097 Manufactured ice
2098 Macaroni, spaghetti, vermicelli, and noodles
2099 Food preparations, n.e.c.*

21 Tobacco Products

2111 Cigarettes
2121 Cigars
2131 Chewing and smoking tobacco and snuff
2141 Tobacco stemming and redrying

22 Textile Mill Products

2211 Broadwoven fabric mills, cotton
2221 Broadwoven fabric mills, manmade fiber, and silk
2231 Broadwoven fabric mills, wool (including dyeing and finishing)
2241 Narrow fabric and other smallwares mills: cotton, wool, silk, and manmade fiber
2251 Women's full length and knee length hosiery, except socks
2252 Hosiery, n.e.c.*
2253 Knit outerwear mills
2254 Knit underwear and nightwear mills
2257 Weft knit fabric mills
2258 Lace and warp knit fabric mills
2259 Knitting mills, n.e.c.*
2261 Finishers of broadwoven fabrics of cotton
2262 Finishers of broadwoven fabrics of manmade fiber and silk
2269 Finishers of textiles, n.e.c.*
2273 Carpets and rugs
2281 Yarn spinning mills
2282 Yarn texturizing, throwing, twisting, and winding mills
2284 Thread mills
2295 Coated fabrics, not rubberized
2296 Tire cord and fabrics
2297 Nonwoven fabrics
2298 Cordage and twine
2299 Textile goods, n.e.c.*

*"Not elsewhere classified" indicated by "*n.e.c.*"

23 Apparel and Other Finished Products made from Fabrics and Other Similar Materials

2311 Men's and boys' suits, coats, and overcoats
2321 Men's and boys' shirts, except work shirts
2322 Men's and boys' underwear and nightwear
2323 Men's and boys' neckwear
2325 Men's and boys' separate trousers and slacks
2326 Men's and boys' work clothing
2329 Men's and boys' clothing, n.e.c.*
2331 Women's, misses', and juniors' blouses and shirts
2335 Women's, misses', and juniors' dresses
2337 Women's, misses', and juniors' suits, skirts, and coats
2339 Women's, misses', and juniors', outerwear, n.e.c.*
2341 Women's, misses', children's, and infants' underwear and nightwear
2342 Brassieres, girdles, and allied garments
2353 Hats, caps, and millinery
2361 Girls', children's and infants' dresses, blouses, and shirts
2369 Girls', children's and infants' outerwear, n.e.c.*
2371 Fur goods
2381 Dress and work gloves, except knit and all leather
2384 Robes and dressing gowns
2385 Waterproof outerwear
2386 Leather and sheep lined clothing
2387 Apparel belts
2389 Apparel and accessories, n.e.c.*
2391 Curtains and draperies
2392 Housefurnishings, except curtains and draperies
2393 Textile bags
2394 Canvas and related products
2395 Pleating, decorative and novelty stitching, and tucking for the trade
2396 Automotive trimmings, apparel findings, and related products
2397 Schiffli machine embroideries
2399 Fabricated textile products, n.e.c.*

24 Lumber and Wood Products, Except Furniture

2411 Logging
2421 Sawmills and planing mills, general
2426 Hardwood dimension and flooring mills
2429 Special product sawmills, n.e.c.*
2431 Millwork
2434 Wood kitchen cabinets
2435 Hardwood veneer and plywood
2436 Softwood veneer and plywood
2439 Structural wood members, n.e.c.*
2441 Nailed and lock corner wood boxes and shook
2448 Wood pallets and skids
2449 Wood containers, n.e.c.*
2451 Mobile homes
2452 Prefabricated wood buildings and components
2491 Wood preserving
2493 Reconstituted wood products
2499 Wood products, n.e.c.*

25 Furniture and Fixtures

2511 Wood household furniture, except upholstered
2512 Wood household furniture, upholstered
2514 Metal household furniture
2515 Mattresses, foundations, and convertible beds
2517 Wood television, radio, phonograph, and sewing machine cabinets
2519 Household furniture, n.e.c.*
2521 Wood office furniture
2522 Office furniture, except wood
2531 Public building and related furniture
2541 Wood office and store fixtures, partitions, shelving, and lockers
2542 Office and store fixtures, partitions, shelving, and lockers, except wood
2591 Drapery hardware and window blinds and shades
2599 Furniture and fixtures, n.e.c.*

26 Paper and Allied Products

2611 Pulp mills
2621 Paper mills
2631 Paperboard mills
2652 Setup paperboard boxes
2653 Corrugated and solid fiber boxes
2655 Fiber cans, tubes, drums, and similar products
2656 Sanitary food containers, except folding
2657 Folding paperboard boxes, including sanitary
2671 Packaging paper and plastics film, coated and laminated
2672 Coated and laminated paper, n.e.c.*
2673 Plastics, foil, and coated paper bags
2674 Uncoated paper and multiwall bags
2675 Die-cut paper and paperboard and cardboard
2676 Sanitary paper products
2677 Envelopes
2678 Stationery tablets, and related products
2679 Converted paper and paperboard products, n.e.c.*

*"Not elsewhere classified" indicated as "*n.e.c.*"

27 Printing, Publishing, and Allied Industries

2711 Newspapers: publishing, or publishing and printing
2721 Periodicals: publishing, or publishing and printing
2731 Books: publishing, or publishing and printing
2732 Book printing
2741 Miscellaneous publishing
2752 Commercial printing, lithographic
2754 Commercial printing, gravure
2759 Commercial printing, n.e.c.*
2761 Manifold business forms
2771 Greeting cards
2782 Blankbooks, looseleaf binders and devices
2789 Bookbinding and related work
2791 Typesetting
2796 Platemaking and related services

28 Chemicals and Allied Products

2812 Alkalies and chlorine
2813 Industrial gases
2816 Inorganic pigments
2819 Industrial inorganic chemicals, n.e.c.*
2821 Plastics materials, synthetic resins, and non-vulcanizable elastomers
2822 Synthetic rubber (vulcanizable elastomers)
2823 Cellulosic manmade fibers
2824 Manmade organic fibers, except cellulosic
2833 Medicinal chemicals and botanical products
2834 Pharmaceutical preparations
2835 In vitro and in vivo diagnostic substances
2836 Biological products, except diagnostic substances
2841 Soap and other detergents, except specialty cleaners
2842 Specialty cleaning, polishing, and sanitation preparations
2843 Surface active agents, finishing agents, sulfonated oils, and assistants
2844 Perfumes, cosmetics, and other toilet preparations
2851 Paints, varnishes, lacquers, enamels, and allied products
2861 Gum and wood chemicals
2865 Cyclic organic crudes and intermediates, and organic dyes and pigments
2869 Industrial organic chemicals, n.e.c.*
2873 Nitrogenous fertilizers
2874 Phosphatic fertilizers
2875 Fertilizers, mixing only
2879 Pesticides and agricultural chemicals, n.e.c.*
2891 Adhesives and sealants
2892 Explosives
2893 Printing ink
2895 Carbon black
2899 Chemicals and chemical preparations, n.e.c.*

29 Petroleum Refining and Related Industries

2911 Petroleum refining
2951 Asphalt paving mixtures and blocks
2952 Asphalt felts and coatings
2992 Lubricating oils and greases
2999 Products of petroleum and coal, n.e.c.*

30 Rubber and Miscellaneous Plastics Products

3011 Tires and inner tubes
3021 Rubber and plastics footwear
3052 Rubber and plastics hose and belting
3053 Gaskets, packing, and sealing devices
3061 Molded, extruded, and lathecut mechanical rubber products
3069 Fabricated rubber products, n.e.c.*
3081 Unsupported plastics film and sheet
3082 Unsupported plastics profile shapes
3083 Laminated plastics plate, sheet, and profile shapes
3084 Plastics pipe
3085 Plastics bottles
3086 Plastics foam products
3087 Custom compounding of purchased plastics resins
3088 Plastics plumbing fixtures
3089 Plastics products, n.e.c.*

31 Leather and Leather Products

3111 Leather tanning and finishing
3131 Boot and shoe cut stock and findings
3142 House slippers
3143 Men's footwear, except athletic
3144 Women's footwear, except athletic
3149 Footwear, except rubber, n.e.c.*
3151 Leather gloves and mittens
3161 Luggage
3171 Women's handbags and purses
3172 Personal leather goods, except women's handbags and purses
3199 Leather goods, n.e.c.*

*"Not elsewhere classified" indicated by "*n.e.c.*"

32 Stone, Clay, Glass and Concrete Products

3211 Flat glass
3221 Glass containers
3229 Pressed and blown glass and glassware, n.e.c.*
3231 Glass products, made of purchased glass
3241 Cement, hydraulic
3251 Brick and structural clay tile
3253 Ceramic wall and floor tile
3255 Clay refractories
3259 Structural clay products, n.e.c.*
3261 Vitreous china plumbing fixtures and china and earthenware fittings and bathroom accessories
3262 Vitreous china table and kitchen articles
3263 Fine earthenware (whiteware) table and kitchen articles
3264 Porcelain electrical supplies
3269 Pottery products, n.e.c.*
3271 Concrete block and brick
3272 Concrete products, except block and brick
3273 Ready mixed concrete
3274 Lime
3275 Gypsum products
3281 Cut stone and stone products
3291 Abrasive products
3292 Asbestos products
3295 Minerals and earths, ground or otherwise treated
3296 Mineral wool
3297 Nonclay refractories
3299 Nonmetallic mineral products, n.e.c.*

33 Primary Metal Industries

3312 Steel works, blast furnaces (including coke ovens), and rolling mills
3313 Electrometallurgical products, except steel
3315 Steel wiredrawing and steel nails and spikes
3316 Cold-rolled steel sheet, strip, and bars
3317 Steel pipe and tubes
3321 Gray and ductile iron foundries
3322 Malleable iron foundries
3324 Steel investment foundries
3325 Steel foundries, n.e.c.*
3331 Primary smelting and refining of copper
3334 Primary production of aluminum
3339 Primary smelting and refining of nonferrous metals, except copper and aluminum
3341 Secondary smelting and refining of nonferrous metals
3351 Rolling, drawing, and extruding of copper
3353 Aluminum sheet, plate, and foil
3354 Aluminum extruded products
3355 Aluminum rolling and drawing, n.e.c.*
3356 Rolling, drawing, and extruding of nonferrous metals, except copper and aluminum
3357 Drawing and insulating of nonferrous wire
3363 Aluminum die-castings
3364 Nonferrous die-castings, except aluminum
3365 Aluminum foundries
3366 Copper foundries
3369 Nonferrous foundries, except aluminum and copper
3398 Metal heat treating
3399 Primary metal products, n.e.c.*

34 Fabricated Metal Products, except Machinery and Transportation Equipment

3411 Metal cans
3412 Metal shipping barrels, drums, kegs, and pails
3421 Cutlery
3423 Hand and edge tools, except machine tools and handsaws
3425 Handsaws and saw blades
3429 Hardware, n.e.c.*
3431 Enameled iron and metal sanitary ware
3432 Plumbing fixture fittings and trim
3433 Heating equipment, except electric and warm air furnaces
3441 Fabricated structural metal
3442 Metal doors, sash, frames, molding, and trim
3443 Fabricated plate work (boiler shops)
3444 Sheet metal work
3446 Architectural and ornamental metal work
3448 Prefabricated metal buildings and components
3449 Miscellaneous structural metal work
3451 Screw machine products
3452 Bolts, nuts, screws, rivets, and washers
3462 Iron and steel forgings
3463 Nonferrous forgings
3465 Automotive stampings
3468 Crowns and closures
3469 Metal stampings, n.e.c.*
3471 Electroplating, plating, polishing, anodizing, and coloring
3479 Coating, engraving and allied services, n.e.c.*
3482 Small arms ammunition
3483 Ammunition, except for small arms
3484 Small arms
3489 Ordnance and accessories, n.e.c.*
3491 Industrial valves
3492 Fluid power valves and hose fittings
3493 Steel springs, except wire
3494 Valves and pipe fittings, n.e.c.*

*"Not elsewhere classified" indicated as "*n.e.c.*"

3495 Wire springs
3496 Miscellaneous fabricated wire products
3497 Metal foil and leaf
3498 Fabricated pipe and pipe fittings
3499 Fabricated metal products, n.e.c.*

35 Industrial and Commercial Machinery and Computer Equipment

3511 Steam, gas and hydraulic turbines, and turbine generator set units
3519 Internal combustion engines, n.e.c.*
3523 Farm machinery and equipment
3524 Lawn and garden tractors and home lawn and garden equipment
3531 Construction machinery and equipment
3532 Mining machinery and equipment, except oil and gas field machinery and equipment
3533 Oil and gas field machinery and equipment
3534 Elevators and moving stairways
3535 Conveyors and conveying equipment
3536 Overhead traveling cranes, hoists, and monorail systems
3537 Industrial trucks, tractors, trailers, and stackers
3541 Machine tools, metal cutting types
3542 Machine tools, metal forming types
3543 Industrial patterns
3544 Special dies and tools, die sets, jigs and fixtures, and industrial molds
3545 Cutting tools, machine tool accessories, and machinists' measuring devices
3546 Power driven handtools
3547 Rolling mill machinery and equipment
3548 Electric and gas welding and soldering equipment
3549 Metalworking machinery, n.e.c.*
3552 Textile machinery
3553 Woodworking machinery
3554 Paper industries machinery
3555 Printing trades machinery and equipment
3556 Food products machinery
3559 Special industry machinery, n.e.c.*
3561 Pumps and pumping equipment
3562 Ball and roller bearings
3563 Air and gas compressors
3564 Industrial and commercial fans and blowers and air purification equipment
3565 Packaging equipment
3566 Speed changers, industrial high speed drives, and gears
3567 Industrial process furnaces and ovens
3568 Mechanical power transmission equipment, n.e.c.*
3569 General industrial machinery and equipment, n.e.c.*
3571 Electronic computers
3572 Computer storage devices
3575 Computer terminals
3577 Computer peripheral equipment, n.e.c.*
3578 Calculating and accounting machines, except electronic computers
3579 Office machines, n.e.c.*
3581 Automatic vending machines
3582 Commercial laundry, drycleaning, and pressing machines
3585 Air conditioning and warm air heating equipment and commercial and industrial refrigeration equipment
3586 Measuring and dispensing pumps
3589 Service industry machinery, n.e.c.*
3592 Carburetors, pistons, piston rings, and valves
3593 Fluid power cylinders and actuators
3594 Fluid power pumps and motors
3596 Scales and balances, except laboratory
3599 Industrial and commercial machinery and equipment, n.e.c*

36 Electronic and Other Electrical Equipment and Components, Except Computer Equipment

3612 Power, distribution, and specialty transformers
3613 Switchgear and switchboard apparatus
3621 Motors and generators
3624 Carbon and graphite products
3625 Relays and industrial controls
3629 Electrical industrial appliances, n.e.c.*
3631 Household cooking equipment
3632 Household refrigerators and home and farm freezers
3633 Household laundry equipment
3634 Electrical housewares and fans
3635 Household vacuum cleaners
3639 Household appliances, n.e.c.*
3641 Electric lampbulbs and tubes
3643 Current carrying wiring devices
3644 Noncurrent carrying wiring devices
3645 Residential electric lighting fixtures
3646 Commercial, industrial, and institutional electric lighting fixtures
3647 Vehicular lighting equipment
3648 Lighting equipment, n.e.c.*
3651 Household audio and video equipment
3652 Phonograph records and pre-recorded audio tapes and disks

*"Not elsewhere classified" indicated by "*n.e.c.*"

3661 Telephone and telegraph apparatus
3663 Radio and television broadcasting and communications equipment
3669 Communications equipment, n.e.c.*
3671 Electron tubes
3672 Printed circuit boards
3674 Semiconductors and related devices
3675 Electronic capacitors
3676 Electronic resistors
3677 Electronic coils, transformers, and other inductors
3678 Electronic connectors
3679 Electronic components, n.e.c.*
3691 Storage batteries
3692 Primary batteries, dry and wet
3694 Electric equipment for internal combustion engines
3695 Magnetic and optical recording media
3699 Electrical machinery, equipment, and supplies, n.e.c.*

37 Transportation Equipment

3711 Motor vehicles and passenger car bodies
3713 Truck and bus bodies
3714 Motor vehicle parts and accessories
3715 Truck trailers
3716 Motor homes
3721 Aircraft
3724 Aircraft engines and engine parts
3728 Aircraft parts and auxiliary equipment, n.e.c.*
3731 Ship building and repairing
3732 Boat building and repairing
3743 Railroad equipment
3751 Motorcycles, bicycles and parts
3761 Guided missiles and space vehicles
3764 Guided missile and space vehicle propulsion units and propulsion unit parts
3769 Guided missile and space vehicle parts and auxiliary equipment, n.e.c.*
3792 Travel trailers and campers
3795 Tanks and tank components
3799 Transportation equipment, n.e.c.*

38 Measuring, Analyzing, and Controlling Instruments; Photographic, Medical and Optical Goods; Watches and Clocks

3812 Search, detection, navigation, guidance, aeronautical, and nautical systems and instruments
3821 Laboratory apparatus and furniture
3822 Automatic controls for regulating residential and commercial environments and appliances
3823 Industrial instruments for measurement, display, and control of process variables; and related products
3824 Totalizing fluid meters and counting devices
3825 Instruments for measuring and testing of electricity and electrical signals
3826 Laboratory analytical instruments
3827 Optical instruments and lenses
3829 Measuring and controlling devices, n.e.c.*
3841 Surgical and medical instruments and apparatus
3842 Orthopedic, prosthetic, and surgical appliances and supplies
3843 Dental equipment and supplies
3844 X-ray apparatus and tubes and related irradiation apparatus
3845 Electromedical and electrotherapeutic apparatus
3851 Ophthalmic goods
3861 Photographic equipment and supplies
3873 Watches, clocks, clockwork operated devices, and parts

39 Miscellaneous Manufacturing Industries

3911 Jewelry, precious metal
3914 Silverware, plated ware, and stainless steel ware
3915 Jewelers' findings and materials, and lapidary work
3931 Musical instruments
3942 Dolls and stuffed toys
3944 Games, toys and children's vehicles; except dolls and bicycles
3949 Sporting and athletic goods, n.e.c.*
3951 Pens, mechanical pencils, and parts
3952 Lead pencils, crayons, and artists' materials
3953 Marking devices
3955 Carbon paper and inked ribbons
3961 Costume jewelry and costume novelties, except precious metal
3965 Fasteners, buttons, needles, and pins
3991 Brooms and brushes
3993 Signs and advertising specialties
3995 Burial caskets
3996 Linoleum, asphalted-felt-base, and other hard surface floor coverings, n.e.c.*
3999 Manufacturing industries, n.e.c.*

*"Not elsewhere classified" indicated as "*n.e.c.*"

Section 313
Toxic Chemical Categories

SECTION 313 TOXIC CHEMICAL CATEGORIES

Section 313 requires reporting on the toxic chemical categories listed below, in addition to the specific toxic chemicals listed above.

The metal compounds listed below, unless otherwise specified, are defined as including any unique chemical substance that contains the named metal (i.e., antimony, copper, etc.) as part of that chemical's structure.

Toxic chemical categories are subject to the 1 percent de minimis concentration unless the substance involved meets the definition of an OSHA carcinogen, which are subject to the 0.1 percent de minimis concentration. The de minimis concentration for each compound is provided in paranthesis.

Antimony Compounds - (Category Code N010) - Includes any unique chemical substance that contains antimony as part of that chemical's infrastructure. (1.0)

Arsenic Compounds - (Category Code N020) - Includes any unique chemical substance that contains arsenic as part of that chemical's infrastructure. (Inorganic compounds: 0.1; organic compounds: 1.0)

Barium Compounds - (Category Code N040) - Includes any unique chemical substance that contains barium as part of that chemical's infrastructure. (1.0)

Beryllium Compounds - (Category Code N050) - Includes any unique chemical substance that contains beryllium as part of that chemical's infrastructure. (Inorganic compounds: 0.1; organic compounds: 1.0)

Cadmium Compounds - (Category Code N078) - Includes any unique chemical substance that contains cadmium as part of that chemical's infrastructure. (Inorganic compounds: 0.1; organic compounds: 1.0)

Chlorophenols - (Category Code N084) - (0.1)

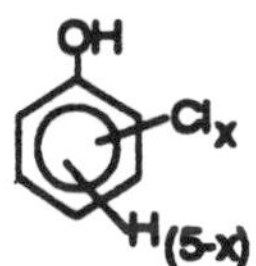

where x = 1 to 5

*C.I. means "Color Index"

Chromium Compounds - (Category Code N090) - Includes any unique chemical substance that contains chromium as part of that chemical's infrastructure. (chromium VI compounds: 0.1; chromium III compounds: 1.0)

Cobalt Compounds - (Category Code N096) - Includes any unique chemical substance that contains cobalt as part of that chemical's infrastructure. (1.0)

Copper Compounds - (Category Code N100) - Includes any unique chemical substance that contains copper as part of that chemical's infrastructure. (1.0)

This category does not include:

Chemical	CAS Number
C.I. Pigment Blue 15	147-14-8
C.I. Pigment Green 7	1328-53-6
C.I. Pigment Green 36	14302-13-7

Cyanide Compounds - (Category Code N106) - $X^+ CN^-$ where $X = H^+$ or any other group where a formal dissociation may occur. For example, KCN or $Ca(CN)_2$. (1.0)

Glycol Ethers - (Category Code N230) - Includes mono- and di- ethers of ethylene glycol, diethylene glycol, and triethylene glycol. (1.0)

$R-(OCH_2CH_2)n-OR'$
Where n = 1,2,or 3

R = alkyl or aryl groups

R'= R, H, or groups which, when removed, yield glycol ethers with the structure:
$R-(OCH_2CH_2)n-OH$

Polymers are excluded from this category.

Lead Compounds - (Category Code N420) - Includes any unique chemical substance that contains lead as part of that chemical's infrastructure. (Inorganic compounds: 0.1; organic compounds: 1.0)

Manganese Compounds - (Category Code N450) - Includes any unique chemical substance that contains manganese as part of that chemical's infrastructure. (1.0)

Mercury Compounds - (Category Code N458) - Includes any unique chemical substance that contains mercury as part of that chemical's infrastructure. (1.0)

Nickel Compounds - (Category Code N495) - Includes any unique chemical substance that contains nickel as part of that chemical's infrastructure. (0.1)

Polybrominated Biphenyls (PBBs) - (Category Code N575) - (0.1)

Br_x
$H_{(10-x)}$

where x = 1 to 10

Selenium Compounds - (Category Code N725) - Includes any unique chemical substance that contains selenium as part of that chemical's infrastructure. (1.0)

Silver Compounds - (Category Code N740) - Includes any unique chemical substance that contains silver as part of that chemical's infrastructure. (1.0)

Thallium Compounds - (Category Code N760) - Includes any unique chemical substance that contains thallium as part of that chemical's infrastructure. (1.0)

Zinc Compounds - (Category Code N982) - Includes any unique chemical substance that contains zinc as part of that chemical's infrastructure. (1.0)

*C.I. means "Color Index"

Toxic Chemical Release Inventory Reporting Form (EPA Form R)

EPA United States Environmental Protection Agency	**FORM R** TOXIC CHEMICAL RELEASE INVENTORY REPORTING FORM Section 313 of the Emergency Planning and Community Right-to-Know Act of 1986, also known as Title III of the Superfund Amendments and Reauthorization Act

TRI FACILITY ID NUMBER

Toxic Chemical, Category, or Generic Name

WHERE TO SEND COMPLETED FORMS:

1. EPCRA Reporting Center
   P.O. Box 23779
   Washington, DC 20026-3779
   ATTN: TOXIC CHEMICAL RELEASE INVENTORY
2. APPROPRIATE STATE OFFICE
   (See instructions in Appendix F)

Enter "X" here if this is a revision

For EPA use only

IMPORTANT: See instructions to determine when "Not Applicable (NA)" boxes should be checked.

PART I. FACILITY IDENTIFICATION INFORMATION

SECTION 1. REPORTING YEAR

19 ____

SECTION 2. TRADE SECRET INFORMATION

2.1	Are you claiming the toxic chemical identified on page 3 trade secret? ☐ Yes (Answer question 2.2; Attach substantiation forms) ☐ No (Do not answer 2.2; Go to Section 3)
2.2	If yes in 2.1, is this copy: ☐ Sanitized ☐ Unsanitized

SECTION 3. CERTIFICATION (Important: Read and sign after completing all form sections.)

I hereby certify that I have reviewed the attached documents and that, to the best of my knowledge and belief, the submitted information is true and complete and that the amounts and values in this report are accurate based on reasonable estimates using data available to the preparers of this report.

Name and official title of owner/operator or senior management official

Signature	Date Signed

SECTION 4. FACILITY IDENTIFICATION

4.1	Facility or Establishment Name	TRI Facility ID Number
	Street Address	
	City	County
	State	Zip Code
	Mailing Address (if different from street address)	PUT LABEL HERE
	City	
	State / Zip Code	

EPA
United States Environmental Protection Agency

EPA FORM R

PART I. FACILITY IDENTIFICATION INFORMATION (CONTINUED)

TRI FACILITY ID NUMBER
Toxic Chemical, Category, or Generic Name

SECTION 4. FACILITY IDENTIFICATION (Continued)

4.2	This report contains information for: **(Important: check only one)**	a. ☐ An entire facility		b. ☐ Part of a facility		
4.3	**Technical Contact**	Name			Telephone Number (include area code)	
4.4	**Public Contact**	Name			Telephone Number (include area code)	
4.5	**SIC Code (4-digit)**	a.	b.	c.	d.	e.	f.

		Latitude			Longitude		
4.6	**Latitude and Longitude**	Degrees	Minutes	Seconds	Degrees	Minutes	Seconds

4.7	**Dun & Bradstreet Number(s) (9 digits)**	a.
		b.
4.8	**EPA Identification Number(s) (RCRA I.D. No.) (12 characters)**	a.
		b.
4.9	**Facility NPDES Permit Number(s) (9 characters)**	a.
		b.
4.10	**Underground Injection Well Code (UIC) I.D. Number(s) (12 digits)**	a.
		b.

SECTION 5. PARENT COMPANY INFORMATION

5.1	Name of Parent Company ☐ **NA**	
5.2	Parent Company's Dun & Bradstreet Number ☐ **NA**	**(9 digits)**

EPA United States Environmental Protection Agency

EPA FORM R

PART II. CHEMICAL-SPECIFIC INFORMATION

TRI FACILITY ID NUMBER
Toxic Chemical, Category, or Generic Name

SECTION 1. TOXIC CHEMICAL IDENTITY

(Important: DO NOT complete this section if you complete Section 2 below.)

1.1	CAS Number (Important: Enter only one number exactly as it appears on the Section 313 list. Enter category code if reporting a chemical category.)
1.2	Toxic Chemical or Chemical Category Name (Important: Enter only one name exactly as it appears on the Section 313 list.)
1.3	Generic Chemical Name (Important: Complete **only** if Part I, Section 2.1 is checked "yes." Generic Name must be structurally descriptive.)

SECTION 2. MIXTURE COMPONENT IDENTITY

(Important: DO NOT complete this section if you complete Section 1 above.)

2.1	Generic Chemical Name Provided by Supplier (Important: Maximum of 70 characters, including numbers,letters, spaces, and punctuation.)

SECTION 3. ACTIVITIES AND USES OF THE TOXIC CHEMICAL AT THE FACILITY

(Important: Check all that apply.)

3.1	Manufacture the toxic chemical:	a. ☐ Produce b. ☐ Import	If produce or import: c. ☐ For on-site use/processing d. ☐ For sale/distribution e. ☐ As a byproduct f. ☐ As an impurity
3.2	Process the toxic chemical:	a. ☐ As a reactant b. ☐ As a formulation component	c. ☐ As an article component d. ☐ Repackaging
3.3	Otherwise use the toxic chemical:	a. ☐ As a chemical processing aid b. ☐ As a manufacturing aid	c. ☐ Ancillary or other use

SECTION 4. MAXIMUM AMOUNT OF THE TOXIC CHEMICAL ON-SITE AT ANY TIME DURING THE CALENDAR YEAR

4.1	☐ (Enter two-digit code from instruction package.)

EPA United States Environmental Protection Agency

EPA FORM R

PART II. CHEMICAL-SPECIFIC INFORMATION (CONTINUED)

TRI FACILITY ID NUMBER
Toxic Chemical, Category, or Generic Name

SECTION 5. RELEASES OF THE TOXIC CHEMICAL TO THE ENVIRONMENT ON-SITE

			A. Total Release (pounds/year) (enter range code from instructions or estimate)	B. Basis of Estimate (enter code)	C. % From Stormwater
5.1	Fugitive or non-point air emissions	☐ NA			
5.2	Stack or point air emissions	☐ NA			
5.3	Discharges to receiving streams or water bodies (enter one name per box)				
5.3.1	Stream or Water Body Name				
5.3.2	Stream or Water Body Name				
5.3.3	Stream or Water Body Name				
5.4	Underground injections on-site	☐ NA			
5.5	Releases to land on-site				
5.5.1	Landfill	☐ NA			
5.5.2	Land treatment/ application farming	☐ NA			
5.5.3	Surface impoundment	☐ NA			
5.5.4	Other disposal	☐ NA			

☐ Check here only if additional Section 5.3 information is provided on page 5 of this form.

Range Codes: A = 1 - 10 pounds; B = 11 - 499 pounds; C = 500 - 999 pounds.

EPA
United States
Environmental Protection
Agency

EPA FORM R

PART II. CHEMICAL-SPECIFIC INFORMATION (CONTINUED)

TRI FACILITY ID NUMBER
Toxic Chemical, Category, or Generic Name

SECTION 5.3 ADDITIONAL INFORMATION ON RELEASES OF THE TOXIC CHEMICAL TO THE ENVIRONMENT ON-SITE

5.3	Discharges to receiving streams or water bodies (enter one name per box)	A. Total Release (pounds/year) (enter range code from instructions or estimate)	B. Basis of Estimate (enter code)	C. % From Stormwater
5.3.__	Stream or Water Body Name			
5.3.__	Stream or Water Body Name			
5.3.__	Stream or Water Body Name			

SECTION 6. TRANSFERS OF THE TOXIC CHEMICAL IN WASTES TO OFF-SITE LOCATIONS

6.1 DISCHARGES TO PUBLICLY OWNED TREATMENT WORKS (POTW)

6.1.A Total Quantity Transferred to POTWs and Basis of Estimate

6.1.A.1 Total Transfers (pounds/year) (enter range code or estimate)	6.1.A.2 Basis of Estimate (enter code)

6.1.B POTW Name and Location Information

6.1.B.__ POTW Name		6.1.B.__ POTW Name	
Street Address		Street Address	
City	County	City	County
State	Zip Code	State	Zip Code

If additional pages of Part II, Sections 5.3 and/or 6.1 are attached, indicate the total number of pages in this box ☐ and indicate which Part II, Sections 5.3/6.1 page this is, here. ☐
(example: 1, 2, 3, etc.)

EPA Form 9350-1 (Rev. 5/14/92) - Previous editions are obsolete.

Range Codes: A = 1 - 10 pounds; B = 11 - 499 pounds; C = 500 - 999 pounds.

EPA

United States Environmental Protection Agency

EPA FORM R

PART II. CHEMICAL-SPECIFIC INFORMATION (CONTINUED)

TRI FACILITY ID NUMBER
Toxic Chemical, Category, or Generic Name

SECTION 6.2 TRANSFERS TO OTHER OFF-SITE LOCATIONS

6.2.___ Off-site EPA Identification Number (RCRA ID No.)

Off-Site Location Name

Street Address

City

County

State

Zip Code

Is location under control of reporting facility or parent company? ☐ Yes ☐ No

A. Total Transfers (pounds/year) (enter range code or estimate)	B. Basis of Estimate (enter code)	C. Type of Waste Treatment/Disposal/ Recycling/Energy Recovery (enter code)
1.	1.	1. M
2.	2.	2. M
3.	3.	3. M
4.	4.	4. M

SECTION 6.2 TRANSFERS TO OTHER OFF-SITE LOCATIONS

6.2.___ Off-site EPA Identification Number (RCRA ID No.)

Off-Site Location Name

Street Address

City

County

State

Zip Code

Is location under control of reporting facility or parent company? ☐ Yes ☐ No

A. Total Transfers (pounds/year) (enter range code or estimate)	B. Basis of Estimate (enter code)	C. Type of Waste Treatment/Disposal/ Recycling/Energy Recovery (enter code)
1.	1.	1. M
2.	2.	2. M
3.	3.	3. M
4.	4.	4. M

If additional pages of Part II, Section 6.2 are attached, indicate the total number of pages in this box ☐ and indicate which Part II, Section 6.2 page this is, here. ☐ (example: 1, 2, 3, etc.)

Range Codes: A = 1 - 10 pounds; B = 11 - 499 pounds; C = 500 - 999 pounds.

EPA United States Environmental Protection Agency

EPA FORM R

PART II. CHEMICAL-SPECIFIC INFORMATION (CONTINUED)

TRI FACILITY ID NUMBER
Toxic Chemical, Category, or Generic Name

SECTION 7A. ON-SITE WASTE TREATMENT METHODS AND EFFICIENCY

☐ Not Applicable (NA) - Check here if no on-site waste treatment is applied to any waste stream containing the toxic chemical or chemical category.

a. General Waste Stream (enter code)	b. Waste Treatment Method(s) Sequence [enter 3-character code(s)]	c. Range of Influent Concentration	d. Waste Treatment Efficiency Estimate	e. Based on Operating Data?
7A.1a	7A.1b 1 ☐ 2 ☐ 3 ☐ 4 ☐ 5 ☐ 6 ☐ 7 ☐ 8 ☐	7A.1c	7A.1d %	7A.1e Yes ☐ No ☐
7A.2a	7A.2b 1 ☐ 2 ☐ 3 ☐ 4 ☐ 5 ☐ 6 ☐ 7 ☐ 8 ☐	7A.2c	7A.2d %	7A.2e Yes ☐ No ☐
7A.3a	7A.3b 1 ☐ 2 ☐ 3 ☐ 4 ☐ 5 ☐ 6 ☐ 7 ☐ 8 ☐	7A.3c	7A.3d %	7A.3e Yes ☐ No ☐
7A.4a	7A.4b 1 ☐ 2 ☐ 3 ☐ 4 ☐ 5 ☐ 6 ☐ 7 ☐ 8 ☐	7A.4c	7A.4d %	7A.4e Yes ☐ No ☐
7A.5a	7A.5b 1 ☐ 2 ☐ 3 ☐ 4 ☐ 5 ☐ 6 ☐ 7 ☐ 8 ☐	7A.5c	7A.5d %	7A.5e Yes ☐ No ☐

If additional copies of page 7 are attached, indicate the total number of pages in this box ☐ and indicate which page 7 this is, here. ☐ (example: 1, 2, 3, etc.)

EPA
United States Environmental Protection Agency

EPA FORM R

PART II. CHEMICAL-SPECIFIC INFORMATION (CONTINUED)

TRI FACILITY ID NUMBER
Toxic Chemical, Category, or Generic Name

SECTION 7B. ON-SITE ENERGY RECOVERY PROCESSES

☐ **Not Applicable (NA)** - **Check here if no on-site energy recovery is applied to any waste stream containing the toxic chemical or chemical category.**

Energy Recovery Methods [enter 3-character code(s)]

1		2		3		4	

SECTION 7C. ON-SITE RECYCLING PROCESSES

☐ **Not Applicable (NA)** - **Check here if no on-site recycling is applied to any waste stream containing the toxic chemical or chemical category.**

Recycling Methods [enter 3-character code(s)]

1		2		3		4		5	
6		7		8		9		10	

EPA
United States
Environmental Protection
Agency

EPA FORM R

PART II. CHEMICAL-SPECIFIC INFORMATION (CONTINUED)

TRI FACILITY ID NUMBER
Chemical, Category, or Generic Name

SECTION 8. SOURCE REDUCTION AND RECYCLING ACTIVITIES

	All quantity estimates can be reported using up to two significant figures.	Column A 1990 (pounds/year)	Column B 1991 (pounds/year)	Column C 1992 (pounds/year)	Column D 1993 (pounds/year)
8.1	Quantity released*				
8.2	Quantity used for energy recovery on-site				
8.3	Quantity used for energy recovery off-site				
8.4	Quantity recycled on-site				
8.5	Quantity recycled off-site				
8.6	Quantity treated on-site				
8.7	Quantity treated off-site				
8.8	Quantity released to the environment as a result of remedial actions, catastrophic events, or one-time events not associated with production processes (pounds/year)				
8.9	Production ratio or activity index				

8.10	Did your facility engage in any source reduction activities for this chemical during the reporting year? If not, enter "NA" in Section 8.10.1 and answer Section 8.11.			
	Source Reduction Activities [enter code(s)]	Methods to Identify Activity (enter codes)		
8.10.1		a.	b.	c.
8.10.2		a.	b.	c.
8.10.3		a.	b.	c.
8.10.4		a.	b.	c.

8.11	Is additional optional information on source reduction, recycling, or pollution control activities included with this report? (Check one box)	YES ☐	NO ☐

* Report releases pursuant to EPCRA Section 329(8) including "any spilling, leaking, pumping, pouring, emitting, emptying, discharging, injecting, escaping, leaching, dumping, or disposing into the environment." Do not include any quantity treated on-site or off-site.

State Section 313 Contacts

STATE DESIGNATED SECTION 313 CONTACTS

Note: **Use the appropriate address for submission of Form R reports to your State.**

Alabama
Mr. E. John Williford, Chief of Operations
Alabama Emergency Response Commission
Alabama Department of Environmental Management
1751 Congressman W.L. Dickinson Drive
Montgomery, AL 36109
(205) 260-2700

Alaska
Ms. Camille Stevens
Alaska State Emergency Response Commission
410 Willoughby, Suite 105
Juneau, AK 99801-1795
(907) 465-5220

American Samoa
Pati Faiai, Director
American Samoa EPA
Office of the Governor
Pago Pago, AS 96799
International Number (684) 633-2304

Arizona
Mr. Carl Funk
Arizona Emergency Response Commission
Division of Emergency Services, Bldg. 341
5636 East McDowell Road
Phoenix, AZ 85008
(602) 231-6326

Arkansas
Mr. John Ward
Depository of Documents
Arkansas Department of Labor
10421 West Markham
Little Rock, AR 72205
(501) 562-7444

California
Mr. Stephen Hanna, Chief
Office of Environmental Information
Californian Environmental Protection Agency
P.O. Box 2815
Sacramento, CA 95812
(916) 324-9924

Colorado
Winfred Bromley
Colorado Emergency Planning Commission
Colorado Department of Health
4210 East 11th Avenue
Denver, CO 80220
(303) 331-4843

Commonwealth of Northern Mariana Islands
Mr. Frank Russell Meecham, III
Division of Environmental Quality
P.O. Box 1304
Saipan, MP 96950
(670) 234-6984

Connecticut
Ms. Sue Vaughn, Title III Coordinator
State Emergency Response Commission
Department of Environmental Protection
State Office Building, Room 146
165 Capitol Avenue
Hartford, CT 06106
(203) 566-4856

Delaware
Mr. Phillip Retallick
Division of Air and Waste Management
Department of Natural Resources and Environmental Control
89 King's Highway
P.O. Box 1401
Dover, DE 19903
(302) 739-4764

District of Columbia
Mr. Stephen E. Rickman
Office of Emergency Preparedness
Frank Reeves Center for Municipal Affairs
2000 14th Street, NW
Washington, DC 20009
(202) 727-6161

Florida
Mr. Jim Loomis
State Emergency Response Commission
Florida Department of Community Affairs
2740 Centerview Drive
Tallahassee, FL 32399-2149
(904) 488-1472
In Florida: 800-635-7179

Georgia
Mr. Burt Langley
Georgia Emergency Response Commission
205 Butler Street, SE
Floyd Tower East
11th Floor, Suite 1166
Atlanta, GA 30334
(404) 656-6905

Guam
Mr. Fred Castro
Guam EPA
D-107 Harmon Plaza
130 Rojas Street
Harmon, GU 96911
(671) 646-8864

Hawaii
Mr. Leslie Au
Hawaii State Emergency Response Commission
Hawaii State Department of Health
P.O. Box 3378
Honolulu, HI 96801-9904
(808) 586-4251

Idaho
Ms. Margaret Ballard
Idaho Emergency Response Commission
State House
Boise, ID 83720
(208) 334-5888

Illinois
Mr. Joe Goodner
Emergency Planning Unit
Office of Emergency Management
Illinois EPA
P.O. Box 19276
2200 Churchill Road
Springfield, IL 62794-9276
(217) 782-3637

Indiana
Mr. Skip Powers
Indiana Emergency Response Commission
5500 West Bradbury Avenue
Indianapolis, IN 46241
(317) 243-5176

Iowa
Mr. Pete Hamlin
Department of Natural Resources
Wallace Building
900 East Grand Avenue
Des Moines, IA 50319
(515) 281-8852

Kansas
Mr. Karl Birns
Right-to-Know Program
Kansas Emergency Response Commission
Mills Building, 5th Floor, Suite 501
109 S.W. 9th Street
Topeka, KS 66612
(913) 296-1690

Kentucky
Ms. Valerie Hudson
Kentucky Department for Environmental Protection
14 Reilly Road
Frankfort, KY 40601
(502) 564-2150

Louisiana
Mr. R. Bruce Hammatt
Emergency Response Coordinator
Department of Environmental Quality
P.O. Box 82215
7290 Bluebonnet
Baton Rouge, LA 70884-2263
(504) 765-0872

Maine
David D. Brown, Chair
State Emergency Response Commission
State House Station Number 72
Augusta, ME 04333
(207) 289-4080
In Maine: (800) 452-8735

Maryland
Ms. Marsha Ways
State Emergency Response Commission
Maryland Department of the Environment
Toxics Information Center
2500 Broening Highway
Baltimore, MD 21224
(301) 631-3800

Massachusetts
Mr. A. David Rodham, Director
Massachusetts EMA
P.O. Box 1496
400 Worcester Rd.
Framingham, Ma. 01701
(508)-820-2000

Michigan
Mr. Kent Kanagey
Title III Coordinator
Michigan Department of Natural Resources
Environmental Response Division
Title III Notification
P.O. Box 30028
Lansing, MI 48909
(517) 373-8481

Minnesota
Mr. Steve Tomlyanovich
Minnesota Emergency Response Commission
175 Bigelow Building
450 North Syndicate
St Paul, MN 55104
(612) 643-3542

Mississippi
Mr. John David Burns
Mississippi Emergency Response Commission
Mississippi Emergency Management Agency
P.O. Box 4501
Jackson, MS 39296-4501
(601) 960-9000

Missouri
Mr. Jim Long
Missouri Emergency Response Commission
Missouri Department of Natural Resources
P.O. Box 3133
Jefferson City, MO 65102
(314) 526-3344

Montana
Mr. Tom Ellerhoff, Co-Chairman
Montana Emergency Response Commission
Environmental Sciences Division
Department of Health & Environmental Sciences
Capitol Station
Cogswell Building A-107
Helena, MT 59620
(406) 444-3948

Nebraska
Mr. John Steinauer, Coordinator
Nebraska Emergency Response Commission
Nebraska Department of Environmental Control
P.O. Box 98922
State House Station
Lincoln, NE 68509-8922
(402) 471-4251

Nevada
Mr. Joseph Quinn, Chief of Operations
State of Nevada, Division of Emergency Management
2525 South Carson Street
Carson City, NV 89710
(702) 687-4240

New Hampshire
Mr. George L. Iverson, Director
New Hampshire State Emergency Management Agency
Title III Program
State Office Park South
107 Pleasant Street
Concord, NH 03301
(603) 271-2231

New Jersey
Mr. Alan Bookman
New Jersey Emergency Response Commission
SARA Title III Section 313
Department of Environmental Protection and Energy
Division of Environmental Quality, Safety, Health, and Analytical Programs
Right-to-Know
Bureau of Hazardous Substances Information
CN-405
Trenton, NJ 08625
(609) 984-5338

New Mexico
Mr. Max Johnson, Title III Coordinator
New Mexico Emergency Response Commission
Chemical Safety Office, Emergency Management Bureau
P.O. Box 1628
Santa Fe, NM 87504-1628
(505) 827-9223

New York
Mr. William Miner
New York Emergency Response Commission
New York State Department Of Environmental Conservation
Bureau of Spill Prevention and Response
50 Wolf Road/Room 326
Albany, NY 12233-3510
(518)457-4107

North Carolina
Ms. Emily Kilpatrick
North Carolina Emergency Response Commission
North Carolina Division of Emergency Management
116 West Jones Street
Raleigh, NC 27603-1335
Attn: Emily Kilpatrick
(919) 733-3865

North Dakota
Mr. Bob Johnston
North Dakota Emergency Response Commission
Division of Emergency Management
P.O. Box 5511
Bismarck, ND 58502-5511
(701) 224-4589

Ohio
Ms. Cindy DeWulf
Ohio EPA
Division of Air Pollution Control
1800 Watermark Drive
Columbus, OH 43215
(614) 644-3604

Oklahoma
Larry Gales
Oklahoma Department of Health
Environmental Health Administration - 0200
1000 N.E. 10th Street
Oklahoma City, OK 73117-1299
(405) 271-8056

Oregon
Mr. Dennis Walthall
Oregon Emergency Response Commission
c/o State Fire Marshall
4760 Portland Road, N.E.
Salem, OR 97305-1760
(503) 378-3473

Pennsylvania
Mr. James Tinney
Pennsylvania Emergency Management Council
Bureau of Worker and Community Right-to-Know
Room 1503
Labor and Industry Building
7th & Foster Streets
Harrisburg, PA 17120
(717) 783-2071

Puerto Rico
Mr. Pedro Maldonado,
Puerto Rico Emergency Response Commissioner
Title III-SARA Section 313
Puerto Rico Environmental Quality Board
Sernades Junco Station
P.O. Box 11488
Santurce, PR 00910
(809) 767-8181

Rhode Island
Ms. Martha Delaney Mulcahey
Rhode Island Department of Environmental Management
Division of Air and Hazardous Materials
291 Promenade Street
Providence, RI 02908-5767
Attn: Toxic Release Inventory
(401) 277-2808

South Carolina
Mr. Michael Juras
South Carolina Department of Health and Environmental Control
2600 Bull Street
Columbia, SC 29201
Attn: EPCRA Reporting
(803) 935-6336

South Dakota
Ms. Lee Ann Smith, Title III Coordinator
South Dakota Emergency Response Commission
South Dakota Department of Environment and Natural Resources
Joe Foss Building
523 East Capitol
Pierre, SD 57501-3151
(605) 773-3296

Tennessee
Mr. Lacy Suiter, Chairman
Tennessee Emergency Response Commission
Director, Tennessee Emergency Management Agency
3041 Sidco Drive
Nashville, TN 37204
(615) 741-0001
1-800-262-3300 (in Tennessee)
1-800-258-3300 (out of state)

Texas
Ms. Becky Kuicka, Supervisor
Office of Pollution Prevention and Conservation
Texas Water Commission
P.O. Box 13087-Capitol Station
Austin, TX 78711-3087
(512) 463-7869

Utah
Mr. Neil Taylor
Utah Hazardous Chemical Emergency Response Commission
Utah Department of Environmental Quality
Division of Environmental Response and Remediation
1950 West North Temple
Salt Lake City, UT 84116-4840
(801) 536-4100

Vermont
Dr. Jan Carney, Commissioner
Department of Health
60 Main Street
P.O. Box 70
Burlington, VT 05402
(802) 863-7281

Virginia
Ms. Sharon Kenneally-Baxter
Virginia Emergency Response Council
c/o Department of Waste Management
James Monroe Building
14th Floor
101 North 14th Street
Richmond, VA 23219
(804) 225-2581

Virgin Islands
Mr. Roy E. Adams, Commissioner
Department of Planning and Natural Resources
U.S. Virgin Islands Emergency Response Commission
Title III
Nisky Center, Suite 231
Charlotte Amalie
St. Thomas, VI 00802
(809) 774-3320/Ext. 101 or 102

Washington
Mr. Idell Hansen, Supervisor
Community Right-To-Know Unit
Department of Ecology
P.O. Box 47659
Olympia, WA 98504-7659
(206) 438-7252

West Virginia
Mr. Carl L. Bradford, Director
West Virginia Emergency Response Commission
West Virginia Office of Emergency Services
Main Capital Building 1, Room EB-80
Charleston, WV 25305
(304) 558-5380

Wisconsin
Department of Natural Resources
P.O. Box 7921
Madison, WI 53707
Attn: Russ Dumst, Toxics Coordinator
(608) 266-9255

Wyoming
Mr. Joseph Daly, Executive Secretary
Wyoming Emergency Response Commission
Wyoming Emergency Management Agency
P.O. Box 1709
Cheyenne, WY 82003
(307) 777-7566

Notes:

(1) If an Indian tribe has chosen to act independently of a state for the purpose of section 313 reporting, facilities located within that Indian community should report to the tribal SERC, or until the SERC is established, the Chief Executive Officer of the Indian tribe, as well as to EPA; (2) Facilities located within the Territories of the Pacific should send a report to the Chief Administrator of the appropriate territory, as well as to EPA.

Sample Supplier Notification Letter

Sample Notification Letter

January 2, 1992

Mr. Edward Burke
Furniture Company of North Carolina
1000 Main Street
Anytown, North Carolina 99999

Dear Mr. Burke:

The purpose of this letter is to inform you that a product that we sell to you, Furniture Lacquer KXZ-1390, contains 20 percent toluene (Chemical Abstracts Service (CAS) number 108-88-3) and 15 percent zinc compounds. We are required to notify you of the presence of toluene and zinc compounds in the product under section 313 of the Emergency Planning and Community Right-to-Know Act of 1986. This law requires certain manufacturers to report on annual emissions of specified toxic chemicals and chemical categories.

If you are unsure whether or not you are subject to the reporting requirements of Section 313, or need more information, call EPA's Emergency Planning and Community Right-To-Know Information Hotline at (800) 535-0202. Your other suppliers should also be notifying you if section 313 toxic chemicals are in the mixtures and trade name products they sell to you.

Please also note that if you repackage or otherwise redistribute this product to industrial customers, a notice similar to this one should be sent to those customers.

Sincerely,

Axel Leaf
Sales Manager
Furniture Products

Sample Supplier Notification Appearing on an MSDS

Sample Notification on an MSDS

Section 313 Supplier Notification

This product contains the following toxic chemicals subject to the reporting requirements of section 313 of the Emergency Planning and Community Right-To-Know Act of 1986 (40 CFR 372):

CAS #	Chemical Name	Percent by Weight
108-88-3	Toluene	20%
NA	Zinc Compounds	15%

This information should be included in all MSDSs that are copied and distributed for this material.

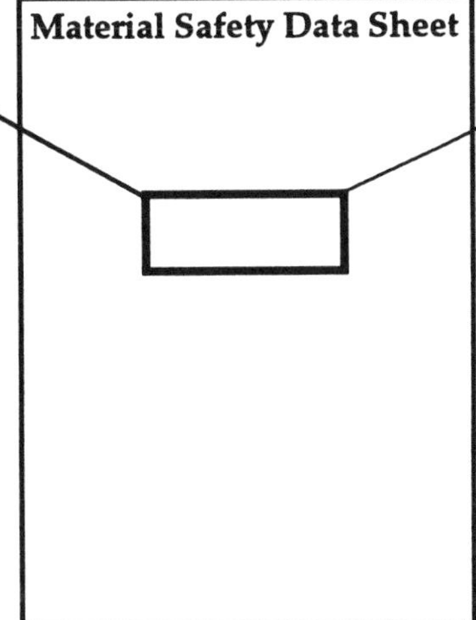

Sample EPA Trade Secret Substantiation Form

United States Environmental Protection Agency
Washington, DC 20460

Substantiation To Accompany Claims of Trade Secrecy Under the Emergency Planning and Community Right-To-Know Act of 1986

Form Approved
OMB No. 2050-0078
Approval expires 02-28-94

Paperwork Reduction Act Notice

Public reporting burden for this collection of information is estimated to vary from 27.7 hours to 33.2 hours per response, with an average of 28.8 hours per response, including time for reviewing instructions, searching existing data sources, gathering and maintaining the data needed, and completing and reviewing the collection of information. Send comments regarding the burden estimate or any other aspect of this collection of information, including suggestions for reducing this burden, to Chief, Information Policy Branch, PM-223, U.S. Environmental Protection Agency, 401 M Street, SW, Washington, DC 20460; and to the Office of Information and Regulatory Affairs, Office of Management and Budget, Washington, DC 20503.

Part 1. Substantiation Category

1.1 Title III Reporting Section (check only one)

☐ 303 ☐ 311 ☐ 312 ☐ 313

1.2 Reporting Year 19________

1.3 Indicate Whether This Form Is (check only one)

1.3a. ☐ **Sanitized**

(answer 1.3.1a below)

1.3.1a. Generic Class or Category

1.3b. ☐ **Unsanitized**

(answer 1.3.1b. and 1.3.2b. below)

1.3.1b. CAS Number

☐☐☐☐☐☐ – ☐☐ – ☐

1.3.2b. Specific Chemical Identity

Part 2. Facility Identification Information

2.1 Name

2.2 Street Address

2.3 City, State, and ZIP Code

2.4 Dun and Bradstreet Number

☐☐ – ☐☐☐ – ☐☐☐☐

Part 3. Responses to Substantiation Questions

3.1 Describe the specific measures you have taken to safeguard the confidentiality of the chemical identity claimed as trade secret, and indicate whether these measures will continue in the future.

3.2 Have you disclosed the information claimed as trade secret to any other person (other than a member of a local emergency planning committee, officer or employee of the United States or a State or local government, or your employee) who is not bound by a confidentiality agreement to refrain from disclosing this trade secret information to others?

☐ Yes ☐ No

3.3 List all local, State, and Federal government entities to which you have disclosed the specific chemical identity. For each, indicate whether you asserted a confidentiality claim for the chemical identity and whether the government entity denied that claim.

Government Entity	Confidentiality Claim Asserted		Confidentiality Claim Denied	
	Yes	No	Yes	No

3.4 In order to show the validity of a trade secrecy claim, you must identify your specific use of the chemical claimed as trade secret and explain why it is a secret of interest to competitors. Therefore:

(i) Describe the specific use of the chemical claimed as trade secret, identifying the product or process in which it is used. (If you use the chemical other than as a component of a product or in a manufacturing process, identify the activity where the chemical is used.)

(ii) Has your company or facility identity been linked to the specific chemical identity claimed as trade secret in a patent, or in publications or other information sources available to the public or your competitors (of which you are aware)?

☐ **Yes** ☐ **No**

If so, explain why this knowledge does not eliminate the justification for trade secrecy.

(iii) If this use of the chemical claimed as trade secret is unknown outside your company, explain how your competitors could deduce this use from disclosure of the chemical identity together with other information on the Title III submittal form.

3.4 (iv) Explain why your use of the chemical claimed as trade secret would be valuable information to your competitors.

3.5 Indicate the nature of the harm to your competitive position that would likely result from disclosure of the specific chemical identity, and indicate why such harm would be substantial.

3.6 (i) To what extent is the chemical claimed as trade secret available to the public or your competitors in products, articles, or environmental releases?

3.6 (ii) Describe the factors which influence the cost of determining the identity of the chemical claimed as trade secret by chemical analysis of the product, article, or waste which contains the chemical (e.g., whether the chemical is in pure form or is mixed with other substances).

Part 4. Certification (Read and sign after completing all sections)

I certify under penalty of law that I have personally examined the information submitted in this and all attached documents. Based on my inquiry of those individuals responsible for obtaining the information, I certify that the submitted information is true, accurate, and complete, and that those portions of the substantiation claimed as confidential would, if disclosed, reveal the chemical identity being claimed as a trade secret, or would reveal other confidential business or trade secret information. I acknowledge that I may be asked by the Environmental Protection Agency to provide further detailed factual substantiation relating to this claim of trade secrecy, and certify to the best of my knowledge and belief that such information is available. I understand that if it is determined by the Administrator of EPA that this trade secret claim is frivolous, EPA may assess a penalty of up to $25,000 per claim.

I acknowledge that any knowingly false or misleading statement may be punishable by fine or imprisonment or both under applicable law.

4.1 Name and official title of owner or operator or senior management official

4.2 Signature *(All signatures must be original)*	4.3 Date Signed

Instructions for Completing the EPA Trade Secret Substantiation Form

General Information

EPA requires that the information requested in a trade secret substantiation be completed using this substantiation form in order to ensure that all facility and chemical identifier information, substantiation questions, and certification statements are completed. Submitter-devised forms will not be accepted. Incomplete substantiations will in all likelihood be found insufficient to support the claim, and the claim will be denied. *Moreover, the statute provides that a submitter who fails to provide information required will be subject to a $10,000 fine.* For the submitter's own protection, therefore, the EPA form must be used and completed in its entirety.

The statute for section 322 establishes a two-phase process in which the submitter must do the following:

1. At the time a report is submitted, the submitter must present a complete set of assertions that (if true) would be sufficient to justify the claim of trade secrecy; and

2. If the claim is reviewed by EPA, the submitter will be asked to provide additional factual information sufficient to establish the truthfulness of the assertions made at the time the claim was made.

In making its assertions of trade secrecy, a submitter should provide, where applicable, descriptive factual statements. Conclusory statements of compliance (such as positive or negative restatements of the questions) may not provide EPA with enough information to make a determination and may be found insufficient to support a claim.

What May Be Withheld

Only the specific chemical identity required to be disclosed in sections 303, 311, 312, and 313 submissions may be claimed trade secret on the Title III submittal itself. (Other trade secret or confidential business information included in answer to a question on the substantiation may be claimed trade secret or confidential, as described below.)

Location information claimed as confidential under section 312(d)(2)(F) should *not* be sent to EPA; this should only be sent to the SERC, LEPC, and the fire department, as requested.

Sanitized and Unsanitized Copies

You must submit this form to EPA in sanitized and unsanitized versions, along with the sanitized and unsanitized copies of the submittal that gives rise to this trade secrecy claim (except for the section 303 submittal, and for MSDSs under section 311). The *unsanitized* version of this form contains specific chemical identity and CAS number and may contain other trade secret or confidential business information, which should be clearly labeled as such. Failure to claim other information trade secret or confidential will make that information publicly available. In the *sanitized* version of this form, the specific chemical identity and CAS number must be replaced with the chemical's generic class or category and any other trade secret or confidential business information should be deleted. *You should also send sanitized copies of the submittal and this form to relevant State and local authorities.*

Each question on this form must be answered. *Submitters are encouraged to answer in the space provided.* If you need more space to answer a particular question, please use additional sheets. If you use additional sheets, be sure to include the number (and if applicable, the subpart) of the question being answered and write your facility's Dun and Bradstreet Number on the lower right-hand corner of each sheet.

When the Forms Must be Submitted

The sanitized and unsanitized report forms and trade secret substantiations must be submitted to EPA by the normal reporting deadline for that section (e.g., section 313 submissions for any calendar year must be submitted on or before July 1 of the following year).

Where to Send the Trade Secrecy Claim

All trade secrecy claims should be sent to the following address: U.S. Environmental Protection Agency, Emergency Planning and Community Right-to-Know Program, P.O. Box 70266, Washington, DC 20024–0266.

In addition, you must send sanitized copies of the report form and substantiation to relevant State and local authorities. States will provide addresses where the copies of the reports are to be sent.

Packaging of Claim(s)

A completed section 322 claim package must include four items, packaged in the following order:

1. An unsanitized trade secret substantiation form.

2. A sanitized trade secret substantiation form.

3. An unsanitized 312 or 313 report (it is not necessary to create an unsanitized section 303 submittal or MSDS for submission under section 311).

4. A sanitized (public) section 303, 311, 312, or 313 or report.

It is important to securely fasten together (binder clip or rubber band) each of the reporting forms and substantiations for the particular chemical being claimed trade secret. This process will make it clear that a claim is physically complete when submitted. When submitters submit claims for more than one chemical, EPA requests that the four parts associated with each chemical be assembled as a set and each set for different chemicals be kept separate within the package sent to EPA. Following these guidelines permits the Agency to make the appropriate determinations of trade secrecy, and to make public only those portions of each submittal required to be disclosed.

How to Obtain Forms and Other Information

Additional copies of the Trade Secret Substantiation Form may be obtained by writing to: Emergency Planning and Community Right-to-Know Program, U.S. Environmental Protection Agency, WH–562A, 401 M Street, SW., Washington, DC 20460.

Instructions for Completing Specific Sections of the Form

Part 1. Substantiation Category

1.1 Title III Reporting Section. Check the box corresponding to the section for which this particular claim of trade secrecy is being made. Checking off more than one box for a claim is *not* permitted.

1.2 Reporting Year. Enter the year to which the reported information applies, not the year in which you are submitting the report.

1.3a Sanitized. If this copy of the submission is the "public" or sanitized version, check this box and complete 1.3.1a. which asks for generic class or category. Do *not* complete the information required in the unsanitized box (1.3b.).

1.3.1a Generic Class or Category. You must complete this if you are claiming the specific chemical identity as a trade secret and have marked the box in 1.3a. The generic chemical name must be structurally descriptive of the chemical.

1.3b Unsanitized. Check the box if this version of the form contains the specific chemical identity or any other trade secret or confidential business information.

1.3.1b CAS Number. You must enter the Chemical Abstract Service (CAS) registry number that appears in the appropriate section of the rule for the chemical being reported. Use leading place holding zeros. If you are reporting a chemical category (e.g., copper compounds), enter N/A in the CAS number space.

1.3.2b Specific Chemical Identity. Enter the name of the chemical or chemical category as it is listed in the appropriate section of the reporting rule.

Part 2. Facility Identification Information

2.1–2.3 Facility Name and Location. You must enter the name of your facility (plant site name or appropriate facility designation), street address, city, State and ZIP Code in the space provided. You may not use a post office box number for this location.

2.4 Dun and Bradstreet Number. You must enter the number assigned by Dun and Bradstreet for your facility or each establishment within your facility. If the establishment does not have a D & B number, enter N/A in the boxes reserved for those numbers. Use leading place holding zeros.

Part 3. Responses to Substantiation Questions

The six questions posed in this form are based on the four statutory criteria found in section 322(b) of Title III. The information you submit in response to these questions is the basis for EPA's initial determination as to whether the substantiation is sufficient to support a claim of trade secrecy. EPA has indicated in § 350.13 of the final rule the specific criteria that it regards as the legal basis for evaluating whether the answers you have provided are sufficient to warrant protection of the chemical identity. You are urged to review those criteria before preparing answers to the questions on the form.

Part 4. Certification

An *original* signature is required for each trade secret substantiation submitted to EPA, both sanitized and unsanitized. It indicates the submitter is certifying that the particular substantiation provided to EPA is complete, true, and accurate, and that it is intended to support the specific trade secret claim being made. Noncompliance with this certification requirement may jeopardize the trade secret claim.

4.1 Name and Official Title. Print or type the name and title of the person who signs the statement at 4.2.

4.2 Signature. This certification must be signed by the owner or operator, or a senior official with management responsibility for the person (or persons) completing the form. An *original* signature is required for each trade secret substantiation submitted to EPA, both sanitized and unsanitized. Since the certification applies to all information supplied on the forms, it should be signed only after the substantiation has been completed.

4.3 Date. Enter the date when the certification was signed.

Appendix A—Restatement of Torts Section 757, Comment b

b. Definition of trade secret. A trade secret may consist of any formula, pattern, device or compilation of information which is used in one's business, and which gives him an opportunity to obtain an advantage over competitors who do not know or use it. It may be a formula for a chemical compound, a process of manufacturing, treating or preserving materials, a pattern for a machine or other device, or a list of customers. It differs from other secret information in a business (see section 759) in that it is not simply information as to single or ephemeral events in the conduct of the business, as, for example, the amount or other terms of a secret bid for a contract or the salary of certain employees, or the security investments made or contemplated, or the date fixed for the announcement of a new policy or for bringing out a new model or the like. A trade secret is a process or device for continuous use in the operation of the business. Generally it relates to the production of goods, as, for example, a machine or formula for the production of an article. It may, however, relate to the sale of goods or to other operations in the business, such as a code for determining discounts, rebates or other concessions in a price list or catalogue, or a list of specialized customers, or a method of bookkeeping or other office management.

Secrecy. The subject matter of a trade secret must be secret. Matters of public knowledge or of general knowledge in an industry cannot be appropriated by one as his secret. Matters which are completely disclosed by the goods which one markets cannot be his secret. Substantially, a trade secret is known only in the particular business in which it is used. It is not requisite that only the proprietor of the business know it. He may, without losing his protection, communicate it to employees involved in its use. He may likewise communicate it to others pledged to secrecy. Others may also know of it independently, as, for example, when they have discovered the process or formula by independent invention and are keeping it secret. Nevertheless, a substantial element of secrecy must exist, so that, except by the use of improper means, there would be difficulty in acquiring the information. An exact definition of a trade secret is not possible. Some factors to be considered in determining whether given information is one's trade secret are: (1) The extent to which the information is known outside of his business; (2) the extent to which it is known by employees and others involved in his business, (3) the extent of measures taken by him to guard the secrecy of the information; (4) the value of the information to him and to his competitors; (5) the amount of effort or money expended by him in developing the information; (6) the ease or difficulty with which the information could be properly acquired or duplicated by others.

Novelty and prior art. A trade secret may be a device or process which is patentable; but it need not be that. It may be a device or process which is clearly anticipated in the prior art or one which is merely a mechanical improvement that a good mechanic can make. Novelty and invention are not requisite for a trade secret as they are for patentability. These requirements are essential to patentability because a patent protects against unlicensed use of the patented device or process even by one who discovers it properly through independent research. The patent monopoly is a reward to the inventor. But such is not the case with a trade secret. Its protection is not based on a policy of rewarding or otherwise encouraging the development of secret processes or devices. The protection is merely against breach of faith and reprehensible means of learning another's secret. For this limited protection it is not appropriate to require also the kind of novelty and invention which is a requisite of patentability. The nature of the secret is, however, an important factor in determining the kind of relief that is appropriate against one who is subject to liability under the rule stated in this section. Thus, if the secret consists of a device or process which is a novel invention, one who acquires the secret wrongfully is ordinarily enjoined from further use of it and is required to account for the profits derived from his past use. If, on the other hand, the secret consists of mechanical improvements that a good mechanic can make without resort to the secret, the wrongdoer's liability may be limited to damages, and an injunction against future use of the improvements made with the aid of the secret may be inappropriate.

EPA Penalty Policy for Sections 302, 303, 304, 311, and 312

OSWER DIR. #9841.2

Final Penalty Policy
for Sections 302, 303, 304, 311, and 312 of the Emergency Planning and Community Right-to-Know Act and Section 103 of the Comprehensive Environmental Response, Compensation and Liability Act

United States Environmental Protection Agency

Office of Solid Waste and Emergency Response
Office of Waste Programs Enforcement

and

Office of Enforcement

[June 13, 1990]

Note: This is a manual reproduction of the original Penalty Policy. Originals can be obtained from EPA Regional Offices.

Table of Contents

I. INTRODUCTION

The Superfund Amendments and Reauthorization Act of 1986 (SARA) created the Emergency Planning and Community Right-To-Know Act (EPCRA). EPCRA §325 authorizes the U.S. EPA Administrator to issue orders compelling owners or operators of facilities to comply with §§302(c) and 303(d) relating to Emergency Planning and to assess penalties administratively for violations of §304 Emergency Notification, §311 Material Safety Data Sheets, §312 Emergency and Hazardous Chemical Inventory, §313 Toxic Chemical Release Forms, §322 Trade Secrets, and §323(b) Provision of Information to Health Professionals, Doctors and Nurses. The EPA Administrator delegated this authority to the Regional Administrators by EPA delegation No. 22-3 dated September 13, 1987. Delegation 22-3 was updated by the Administrator on October 31, 1989.

SARA also amended the enforcement provisions for violation of §103(a) or (b) of the Comprehensive Environmental Response, Compensation, and Liability Act (CERCLA). CERCLA §103(a) and (b) require the person in charge of a facility or vessel to notify the National Response Center (NRC) immediately after the release of a hazardous substance in an amount that exceeds its reportable quantity. CERCLA §109 authorizes the President to assess penalties for violations of CERCLA §103(a) and (b). This authority has since been delegated to the Regional Administrators through the EPA Administrator by EPA delegation No. 14-31 dated September 13, 1987.

Because the reporting requirements for CERCLA §103(a) and (b) and EPCRA §304 are similar and violations of these provisions may arise out of the same set of facts, EPA has decided to combine enforcement of these provisions where possible. Also, EPA proposed in a formal rulemaking that all EPCRA extremely hazardous substances (EHSs) be included on the CERCLA hazardous substance list. When this is accomplished, releases required to be reported under EPCRA §304 will require notification of the National Response Center (NRC) as well.

This penalty policy will provide guidance to Regional EPA case development teams in assessing administrative penalties for violations of CERCLA §103(a) and (b) and EPCRA §§304, 311, and 312. This policy should also be used to develop internal negotiation penalty figures for civil judicial enforcement actions. Other EPCRA provisions not covered by this policy include §§313, 322, and 323. On December 2, 1988, the Office of Pesticides and Toxic Substances (OPTS) issued penalty assessment guidance for violations of §313. The Office of Waste Programs Enforcement (OWPE) is coordinating with WPTS to develop an enforcement response policy for EPCRA §§322 and 323.

This penalty policy provides a framework for assessing penalties within the Agency's established goals of:

- o fair and equitable enforcement of the regulated community and appropriateness of the penalty to the gravity of the violation committed;
- o deterrence; and
- o swift resolution of environmental problems.

This policy does not discuss whether or not an enforcement action seeking a penalty is the correct enforcement response given the specific violative condition. Rather, this policy focuses on determining what the proper civil penalty should be given that a decision has been made to pursue that line of enforcement.

This policy is immediately applicable and should be used to calculate penalties for all administrative actions concerning violations of CERCLA §103(a) and (b) and violations of EPCRA §§302, 303, 304, 311, and 312 instituted after the date of the policy, regardless of the date of violation.

In civil judicial cases, EPA may use the policy to calculate the minimum acceptable penalty amount for settlement purposes, and may use the narrative penalty assessment criteria set forth in the policy to argue for as high a penalty as the facts of a case justify. EPA will revise these calculations as the case progresses to the extent new facts arise which warrant different evaluation of the penalty policy criteria.

Because this policy is intended to provide guidance in assessing administrative and civil judicial penalties only, it does not constitute a statement of EPA policy regarding the appropriate circumstances in which the United States may prosecute violations of CERCLA §103 and EPCRA §304, nor the criminal sentence that a Court should impose upon conviction for violations of either of these two provisions of Federal law.

The procedures set out in this document are intended solely for the use of government personnel. They are not intended and cannot be relied upon to create rights, substantive or procedural, enforceable by any party in litigation with the United States. The Agency reserves the right to act at variance with this policy and to change it at any time without public notice.

II. STATUTORY REQUIREMENTS FOR ASSESSING ADMINISTRATIVE PENALTIES UNDER CERCLA §109 AND EPCRA §325.

CERCLA §103(a) and (b) require the person in charge of a facility or vessel from which a CERCLA hazardous substance has been released in an amount that meets or exceeds its reportable quantity (RQ) to immediately notify the NRC as soon as he/she has knowledge of the release. Violation of the requirements of CERCLA §103 may result in a Class I penalty not to exceed $25,000 per violation. CERCLA §109(a)(3) states that in assessing a Class I penalty for violations of CERCLA §103, EPA must take into account the "nature, circumstances, extent and gravity of the violation or violations and, with respect to the violator, ability to pay, any prior history of such violations, the degree of culpability, economic benefit or savings (if any) resulting from the violation, and other matters as justice may require."

Violations of CERCLA §103(a) or (b) may also result in a Class II penalty not to exceed $25,000 per day for each day the violation continues. For second or subsequent violations, the amount of the Class II penalty is not to exceed $75,000 for each day in which the violation continues. CERCLA §109(b) states that Class II penalties shall be assessed and collected in the same manner, and subject to the some provisions, as in the case of civil penalties assessed and collected after notice and opportunity for hearing on the record in accordance with the Administrative Procedure Act, 5 U.S.C. 554.

Under CERCLA §103(b)(3), any person who fails to notify the appropriate agency of the United States Government or who submits in such notification any information which he knows to be false and misleading shall, upon conviction, be fined in accordance with the applicable provisions of Title 18 of the U.S. Code or imprisoned for not more than 3 years (or not more than 5 years for a second or subsequent conviction), or both.

EPCRA §302 requires the owner or operator of a facility that has present any extremely hazardous substances (EHSs) in amounts that exceed the chemical-specific threshold planning quantity (TPQ) to notify the State Emergency Response Commission (SERC) that the facility is subject to the planning provisions of the Act. For facilities with existing inventories of EHSs in excess of the TPQs, the deadline for notification was May 17, 1987. Thereafter, if a facility newly acquires and EHS in excess of the TPQ, the owner or operator is required to notify the SERC and the Local Emergency Planning Committee (LEPC) within 60 days. EPCRA §325(a) authorizes EPA to issue orders compelling compliance. The U.S. District Court has authority to enforce the order and assess penalties of up to $25,000 per violation per day.

Section 303(d) requires owners or operators subject to §302 to provide the LEPC with the name of a person who will act as the facility emergency coordinator. Additionally, §303(d)(3) requires the owner or operator to promptly supply information to the LEPC upon request. The scope of the information request encompasses anything necessary for developing and implementing the emergency plan. EPA is authorized to issue orders compelling compliance with §303(d). The U.S District Court has authority to enforce the order and assess penalties of up to $25,000 per violation per day.

EPCRA §304(a) requires the owner or operator to notify immediately the appropriate governmental entities for any release that requires CERCLA notification and for releases of EPCRA §302 EHSs. The notification must be given to the SERCs for all States affected by the release and to the community emergency coordinators for the LEPCs for all areas affected by the release. Additionally, EPCRA §304(c) requires any owner or operator who has had a release that is reportable under EPCRA §304(a) to provide, as soon as practicable, a follow-up written notice (or notices) updating the information required under §304(b).

Section 325(b)(1) authorizes EPA to assess a Class I penalty of up to $25,000 per violation of any requirement of §304. EPCRA §325(b)(1)(c) states that in determining the amount of any Class I penalty assessed for a violation of §304, the Administrator shall take into account the "nature, circumstances, extent and gravity of the violation or violations and, with respect to the violator, ability to pay, any prior history of such violations, the degree of culpability, economic benefit or savings (if any) resulting from the violation, and other matters as justice may require."

Section 325(b)(2) authorizes the Administrator to assess a Class II penalty for violations of §304 in an amount not to exceed $25,000 for each day a violation continues. For second or subsequent violations, the amount of the Class II penalty is not to exceed $75,000 for each day in which the violation continues. Any civil penalty under §325(b)(2) shall be assessed and collected in the same manner, and subject to the same provisions as in the case of civil penalties assessed and collected under §16 of the Toxic Substances Control Act (TSCA). TSCA §16 mandates that EPA consider the same factors in assessing penalties that are laid out in EPCRA §325(b)(1)(C) and includes the additional requirement for EPA to consider the effect on the ability to continue to do business. EPA interprets EPCRA §325(b)(2) to mean that the Agency must follow the procedural aspects of TSCA §16 (i.e., using the Consolidated Rules of Practice codified at 40 CFR Part 22) and consider §16 statutory factors for assessing penalties, but not any specific penalty policies developed by the Agency under TSCA §16.

Under EPCRA §325(b)(4), any person who knowingly and willfully fails to provide notice in accordance with section 304, shall, upon conviction, be fined not more than $25,000 or imprisoned not more than two years, or both. In the case of a second or subsequent conviction, such person shall be fined not more than $50,000 or imprisoned for not more than 5 years, or both.

EPCRA §311 requires that the owner or operator of a facility who is required to prepare or have available a Material Safety Data Sheet (MSDS) for a hazardous chemical under the Occupational Safety and Health Act of 1970, shall submit to the SERC, LEPC, and the fire department with jurisdiction over the facility, on or before October 17, 1987 (or 3 months after the owner or operator first becomes subject to OSHA), a MSDS for each such chemical present at the facility in quantities equal to or greater than 10,000 pounds (or a list of such chemicals as described in that section). If the hazardous chemical is a listed EHS under §302, the threshold for reporting is 500 pounds or the chemical-specific threshold planning quantity (TPQ), whichever is less.

For facilities newly covered by EPCRA §311 as a result of the OSHA Hazard Communication Standard expansion to the non-manufacturing sector, MSDSs or a list of MSDSs were required to be submitted by September 24, 1988 to the SERC, LEPC and fire department with jurisdiction over the facility. Construction industry facilities, due to a court ordered delay, were required to comply by April 30, 1989. Additionally, if a facility changes its inventory and becomes subject to EPCRA §311, the facility must report within 3 months.

Section 312 of EPCRA provides that the owner or operator of a facility required to prepare or have available a MSDS for a hazardous chemical under OSHA, shall submit to the SERC, LEPC, and the fire department with jurisdiction over the facility, by March 1, 1988 (and thereafter annually), a completed emergency and hazardous chemical inventory form containing the information required under that section.

For non-manufacturing facilities that were newly covered by the OSHA expansion, the first reporting deadline for EPCRA §312 was March 1, 1989. Facilities in the construction industry must report for the first time by March 1, 1990. Additionally, if a facility changes its inventory and becomes subject to EPCRA §312, the facility must report by March 1 of the following year for the previous year's inventory.

EPCRA §325(c) states that any person who violates §§312 is liable for a penalty in an amount not to exceed $25,000 for each violation. For violations of §311, §325(c)(2) provides that the violator is subject to a penalty in an amount not to exceed $10,000 per violation. Section 325(c)(3) states that each day a violation of §§311 or 312 continues constitutes a separate violation. The statute provides no further guidance for calculating penalties under §325(c) for violations of §§311 and 312. However, as a matter of policy, the Agency will use the statutory factors listed in §325(b)(1)(C) as guidance in calculating penalties for §§311 and 312.

III. ELEMENTS OF THE CIVIL PENALTY SYSTEM

The civil penalty system established in this document is designed to comport with the requirements for assessing administrative penalties established in CERCLA §109, 42 U.S.C. 9609 AND EPCRA §325, 42 U.S.C. 11045. Penalties are to be determined in two stages. First, a preliminary deterrence (base) penalty is calculated using the statutory factors1 that apply to the violation (nature, circumstances, extent, and gravity). After that base penalty is calculated, the statutory factors that apply to the violator are considered (ability to pay/continue in business, prior history of violations, the degree of culpability, economic benefit or savings, and other matters as justice may require). Together, the two calculations will yield a penalty amount that considers all the statutory factors and is appropriate for the violation. To determine the base penalty, the following factors related to a violation are considered:

- The "Nature" of the violation;
- The "Extent" of the violation;
- the "Gravity" of the violation; and
- The "Circumstances" of the violation.

These factors are incorporated into one matrix for CERCLA §103 and EPCRA §302, §303, §304, and §312 violations and another

---

[1] Note "statutory factors" apply as a matter of law only to Class I penalties under CERCLA §109(a)(3) for violations of CERCLA 103 and EPCRA §325(b)(1)(C) for violations of §304 only. EPA applies them to Class II penalties and violations of EPCRA §§302, 303, 311, and 312 as a matter of policy.

matrix for §311 violations. Two matrices are used because of the difference in the statutory maximum associcated with the different violations. For §311 the maximum daily amount is $10,000; for CERCLA §103 and EPCRA §§302, 303, 304, and §312, the first violation maximum daily amount is $25,000. The base penalty can be calculated form the matrices in Table I (Infra, Page 20).

The penalty amounts in the matrix were established so that a worst-case scenario violation could result in the statutory maximum penalty being assessed. The Agency believes that the amount of the chemical involved in the violation and the timeliness of the required reports are both significant factors in determining the appropriateness of a penalty. The penalty calculation scheme in this policy assumes that the greater the quantity of chemicals one uses in conducting business operations, the more likely that a violation of the reporting requirements will undermine the emergency planning, emergency response, and right-to-know intentions of CERCLA §103 and EPCRA. It also assumes that the greater the quantity of chemicals one uses in conducting business operations, the responsible one should be for the safe handling of those chemicals in both emergency and non-emergency situations. Thus, the penalty scheme in this policy is equitable and should provide an appropriate level of deterrence to would-be violators.

The two primary factors used to establish the penalty amount in the matrix (gravity and extent) are equally weighted. Thus, the matrices are symmetrical. The penalties range from 100% of the statutory maximum penalty to 10%. The mid range penalty cells reflect 66%, 33%, 25%, and 18% of the statutory maximum penalty. Two penalty amounts are displayed in each cell of the matrices, with the lower amount being 80% of the upper range of the cell. These penalty amounts were developed under the assumption that the violator has the ability to pay.

A. Use of the Matrix

The success of EPCRA is attained primarily through voluntary, strict and comprehensive compliance with the Act and its regulations. Deviation from the reporting requirements weakens the expressed intent of the Act to allow communities to plan for and respond to chemical emergencies and to allow citizens guaranteed access to information on chemical hazards present in their community.

Owners or operators of facilities are responsible for ensuring that reports for each chemical required to be submitted under §§304, 311, and 312 are submitted to all recipients on or before the required deadline. Failure to submit the required report to any one of the recipients by the reporting deadline is a violation subject to the full penalty allowed by the applicable section of EPCRA. For example, if Company X fails to submit a §312 inventory form for its hazardous chemicals to all three points of compliance (SERC, LEPC, fire department), it is liable for three separate $25,000 per day penalties. The term "points of compliance" refers to the specific entities designated to receive submissions and notices under EPCRA.

In assuring penalties, EPA shall consider Respondent's failure to submit required reports to each point of compliance as separate violations. Accordingly, the matrix should be used for each separate violation of a given section. A facility may submit information on one chemical to each of the three recipients at different times. Therefore, the extent factor may be different for each violation resulting in different penalty amounts. For first time violators, where the facts and circumstances of the case warrant it, e.g., it is clear that the respondent had no prior actual knowledge of the CERCLA §103 or EPCRA reporting requirements, Regions have the option to assess one penalty for multi-point of compliance situations. For second or subsequent violations, penalties should be assessed per point of compliance.

The penalty amounts shown in Table I are meant for first time offenders. For second or subsequent violations of CERCLA §103 and EPCRA §304, the Acts authorize penalties of up to $75,000 per violation per day. For these violations, treble the amount shown at the appropriate position of the matrix. Second and subsequent violations of §§311 and 312 may be addressed through per day assessment of the base penalty found on the matrix. (See also, the sections on multi-day penalties and prior history of violations.)

IV. DETERMINATION OF THE BASE PENALTY

A. Nature

Nature describes the type of violation or requirement violated. In the context of this penalty policy, it is used to determine which specific penalty guidelines should be used to determine appropriate matrix levels of extent and gravity. For the purposes of the EPCRA and CERCLA reporting requirements, there are basically two types of violations: emergency response violations and emergency preparedness/right-to-know violations. Emergency response violations are those in which the violator failed to perform some function or duty during or after a release of a CERCLA hazardous substance or an EPCRA extremely hazardous substance (EHS). Emergency preparedness/right-to-know violations are those in which the violator was required to submit notifications, information, or reports under the EPCRA statutory or regulatory timeframes. The types of violations addressed by this penalty policy include, but are not limited to:

i). Emergency Response Violations

- Failure to notify the National Response Center as required under CERCLA §103(a) or failure to provide all of the information required by statute or implementing regulations.

- Failure to notify all affected State Emergency Response Commissions (SERCs) and the emergency response coordinators for all affected local emergency planning committees (LEPCs) immediately as required under §304(a) or failure to provide all of the information required by statute or implementing regulations.

- Failure to submit a written follow-up report to all affected State emergency response commissions (SERCs) and the emergency response coordinators for all affected local emergency planning committees (LEPCs) as soon as practicable after the release as required under §304(c) or failure to provide all of the information required by statute or implementing regulations.

ii). Emergency Preparedness/Right-to-Know Violations

- Failure to notify the State Emergency Response Commission that the facility is subject to the provisions of the Act as required under EPCRA §302.

- Failure to inform the LEPC of the name of the facility emergency coordinator as required under EPCRA §303(d)(1).

- Failure to notify the LEPC of any relevant changes at the facility as required under §303(d)(2).

- Failure to provide information to the Local Emergency Planning Committee (LEPC) upon request as required under EPCRA §303(d)(3).

- Failure to submit Material Safety Data Sheets (MSDSs) or a list of MSDSs (or failure to include a chemical on the list) to each of the following: the appropriate LEPC; the SERC; and the fire department with jurisdiction over the facility as required under EPCRA §311(a).

- Failure to submit a MSDS to the LEPC upon request as required under EPCRA §311(c).

- o Failure to submit (or incomplete submission of) Emergency and a Hazardous Chemical Inventory Form to each of the following: the appropriate LEPC; the SERC; and the fire department with jurisdiction over the facility as required under EPCRA §312.

- o Failure to provide Tier II information as described in EPCRA §312(d) to a SERC, LEPC, or fire department upon request as required under EPCRA §312(e).

B. Extent

The extent factor is used in this penalty policy to reflect the amount of deviation from CERCLA or EPCRA and their regulatory requirements. In other words, a violation may range from being substantially in compliance with the provisions of CERCLA §103 or EPCRA to being in total disregard of the requirement. Because the immediate notification requirements under CERCLA §103 and EPCRA §304 simply require phone calls to the NRC, SERC, and LEPC community emergency coordinator, deviation from the requirement is mainly measured in terms of timeliness. The person providing the notice must provide the information required in 40 CFR Part 355.40 (chemical identity, estimated quantity released, time/duration of the release, etc.) to the extent known at the time of notice and so long as no delay in notice or emergency response results.

LEVEL 1: The violation deviates from the requirements of the statute to such an extent that there is substantial noncompliance.

LEVEL 2: The violation significantly deviates from the requirements of the statute, but some of the requirements are met.

LEVEL 3: The violation deviates somewhat from the requirements of the statute, but there is substantial compliance.

i). Emergency Response Violations

Under both CERCLA and EPCRA, in the event of a reportable release, notification of the proper authorities is required to occur immediately after the owner, operator or person in charge has knowledge of the release. The statutes, and regulations codified at 40 CFR Parts 302 and 355, identify the information required to be reported in the event of an accidental release. A delay in the notification, or incomplete notification, could seriously hamper Federal and State response activities and pose serious threats to human health and the environment. Thus, the extent factor focuses on the notification and follow-up actions taken by the respondent and the expediency with which those actions were taken.

The statutes require that notification be made by the person in charge (or owner or operator) immediately after he/she has knowledge of a release of an RQ or more of a substance. Although this policy does not define "immediate", it does establish guidelines to assist Agency personnel in determining whether or not an "immediate" standard was met. The immediate notification is required to allow Federal, State, and local agencies to determine what level of government response is needed and with what urgency the response must take place. Early and effective communication of the release event is crucial. At some point, the delay in notification is the same as no notification at all. For both the CERCLA and EPCRA notification requirements, EPA may assess the statutory maximum for any notification that does not occur "immediately" after the "person in charge" (CERCLA) or the "owner or operator" (EPCRA) has knowledge of the release. The levels identified below reflect the benefit of expeditious notification by discounting from the maximum statutory penalty for the timeliness of the notification.

LEVEL 1:

CERCLA §103: No notification to the NRC within 2 hours after the person in charge had knowledge that a reportable quantity of a substance was released unless extenuating circumstances existed that prevented notification.

EPCRA §304(a): No notification to the appropriate SERC(s) and LEPC(s) within 2 hours after the owner or operator had knowledge of the release unless extenuating circumstances existed that prevented notification.

EPCRA §304(c): No written follow-up report to the appropriate SERC(s) and LEPC(s) within 2 weeks following the release unless extenuating circumstances prevented its submission.

LEVEL 2:

CERCLA §103: No notification to the NRC within 1 hour (but within 2 hours) after the person in charge had knowledge that a reportable quantity of a substance was released unless extenuating circumstances prevented the notification.

EPCRA §304(a): No notification to the appropriate SERC(s) and LEPC(s) within 1 hour (but within 2 hours) after the owner or operator had knowledge of the release unless extenuating circumstances prevented the notification.

EPCRA §304(c): No written follow-up report to the appropriate SERC(s) and LEPC(s) within 1 week (but within 2 weeks) following the release unless extenuating circumstances prevented its submission.

LEVEL 3:

CERCLA §103: No immediate notification to the NRC, i.e., although notification occurred within one hour, the facts and circumstances indicate that the notification could have been made sooner than actually made.

EPCRA §304(a): No immediate notification to the appropriate SERC(s) and LEPC(s), i.e., although notification occurred within one hour, the facts and circumstances of the incident indicate that the notification(s) could have been made sooner than actually made.

EPCRA §304(c): No written follow-up report to the appropriate SERC(s) and LEPC(s) as soon as practicable, i.e., although follow-up notification occurred within one week, the facts and circumstances of the incident indicate that the follow-up was not as soon as practicable.

ii). Emergency Preparedness/Right-to-know Violations

The emergency preparedness/right-to-know provisions require that owners or operators submit information to State and local entities. For emergency preparedness/right-to-know violations, the extent factor reflects the potential deleterious effect the noncompliance has on the Agency's, SERC's or LEPC's ability to implement the Act or the public's ability to access the information. For each of these violations, the Agency could assess the statutory maximum for each violation on a per day basis. However, the Agency can exercise discretion in penalizing violations to set amount levels below the statutory maximum for differences in the extent of the violation. Extent addresses the timeliness and utility of reports submitted. Therefore, the extent factor is used, in part, to provide some built-in incentives for nonreporters to submit the required reports (self confess) as soon as possible, albeit late, and to provide incentives for submitters to fill out the forms in a manner consistent with the statutory and regulatory requirements.

The goal of this part is to establish a standard for timeliness and completeness. It will allow potential violators to know by what standard penalties may be assessed should they violate EPCRA. It will also promote Agency consistency in assessing penalties by establishing uniform assessments for late reporting and failure to report.

The matrix levels for measuring extent for the emergency planning/right-to-know violations are as follows:

LEVEL 1:

EPCRA §302: Respondent fails to notify the SERC that it is subject to the Act within 30 calendar days of the reporting deadline.

EPCRA §303: Respondent fails to notify the LEPC within 30 calendar days of reporting obligation.

Respondent fails to respond to Administrative Order for §303(d)(3) within 30 calendar days of required response date.

Respondent submits information in response to §303 information request, claims trade secret any chemical identity, but fails to submit trade secret substantiation to justify the claim (thereby rendering the §303 submission substantively incomplete and potentially fraudulent).

EPCRA §311: Respondent fails to submit MSDS for each required hazardous chemical (or list of MSDSs) as required by §311(a) to the SERC, LEPC, or fire department within 30 calendar days of the reporting obligation.

Respondent fails to include chemical on list submitted.

Respondent submits MSDS or list claiming chemical identity a trade secret, but fails to submit trade secret substantiation to justify the claim (thereby rendering the §311 submission substantively incomplete and potentially fraudulent).

Respondent fails to respond to request under §311(c) within 30 calendar days of the reporting obligation.

EPCRA §312: Respondent fails to submit Inventory Form to the SERC, LEPC, or fire department within 30 calendar days of reporting deadline.

Inventory form submitted fails to address each hazard category present at the facility.

Respondent fails to respond to request under §312(e) within 30 calendar days of the reporting obligation.

Respondent submits form that claims trade secret status for chemical identification, but Respondent fails to submit trade secret substantiation to justify the claim (thereby rendering the §312 submission substantively incomplete and potentially fraudulent).

LEVEL 2:

EPCRA §302: Respondent fails to notify the SERC that it is subject to the Act within 20 (but does within 30) calendar days of reporting obligation.

EPCRA §303: Respondent fails to notify the LEPC within 20 (but does within 30) calendar days of reporting obligation.

Respondent fails to respond to an Administrative Order within 20 (but does within 30) calendar days of required response date.

EPCRA §311: Respondent fails to submit MSDS (or list of MSDSs) to the SERC, LEPC, or fire department within 20 (but does within 30) calendar days of reporting obligation.

Respondent fails to respond to request under §311(c) within 20 (but does within 30) calendar days of the reporting obligation.

EPCRA §312: Respondent fails to submit Inventory Form to the SERC, LEPC, or fire department within 20 (but does within 30) calendar days of reporting deadline.

Inventory form submitted covers all hazard categories present at the facility, but fails to cover all hazardous chemicals present at the facility during the preceding calendar year in amounts equal to or greater than the reporting thresholds. Respondent's failure to address all of the hazardous chemicals renders the submission incomplete (i.e., all general locations not supplied) or inaccurate (i.e., different ranges apply).

Respondent fails to respond to request under §312(e) within 20 (but does within 30) calendar days of required response date.

LEVEL 3:

EPCRA §302: Respondent fails to notify the SERC within 10 (but does within 20) calendar days of reporting obligation.

EPCRA §303: Respondent fails to notify the LEPC within 10 (but does with 20) calendar days of reporting obligation.

Respondent fails to respond to an Administrative Order within 10 (but does with 20) calendar days of required response date.

EPCRA §311: Respondent fails to submit MSDS (or list of MSDSs) to the SERC, LEPC, or fire department within 10 (but does within 20) calendar days of reporting obligation.

Respondent fails to respond to request under §311(c) within 10 (but does within 20) calendar days of the reporting obligation.

EPCRA §312: Respondent fails to submit Inventory Form to the SERC, LEPC, or fire department within 10 (but does within 20) calendar days of reporting deadline.

Respondent submitted form addresses all hazard categories, but fails to meet the standard required by the Statute or Rule.

Respondent fails to respond to request under §312(e) within 10 (but does within 20) calendar days of required response date.

C. Gravity

For the purposes of the emergency response violations, gravity is determined by the amount of the substance involved in the violation. CERCLA hazardous substances and EPCRA EHSs have reportable quantities (RQs) that vary depending on the substance, but range from 1 pound to 10,000 pounds. Reportable quantities were established for hazardous substances to indicate an amount, which if exceeded in a release, would require immediate notification to the proper governmental authorities. The RQ scale itself is a relative measure of the hazards posed by the chemical and therefore the potential threat to human health and the environment; the lower the RQ, the greater the potential threat to human health and the environment. The greater the amount released over the RQ, the greater the potential for the need for immediate notification. Likewise, the greater the amount stored on site, the greater the need for fire departments and emergency planners to know of its existence and location prior to any explosion or unpermitted release. The goal of setting standards for the gravity component is to establish, prospectively, the Agency's expectations for those who handle hazardous and extremely hazardous chemicals.

i). Emergency Response Violations

For emergency response violations, the Agency will penalize a failure to notify relative, in part, to the amount by which the RQ was exceeded. To determine gravity for emergency response violations, use the following levels:

LEVEL A: The amount released was greater than 10 times the RQ;

LEVEL B: The amount released was greater than 5, but less than or equal to 10 times the RQ;

LEVEL C: The amount released was greater than 1, but less than or equal to 5 times the RQ.

ii). Emergency Preparedness/Right-to-know Violations

For the purposes of emergency preparedness/right-to-know violations, the number and/or amount of the chemical(s) in excess of the reporting threshold present at the facility forms the basis for determining gravity. For §§311 and 312, the reporting threshold for EHSs is 500 pounds or the EHS-specific threshold planning quantity (TPQ), whichever is less. For other hazardous chemicals, the reporting threshold is 10,000 pounds. Under §311, a MSDS is required for each chemical over the threshold. Alternatively, if a list is submitted, each chemical that exceeded the threshold must be specifically identified on the list. For §311 violations, the gravity levels are:

LEVEL A: Amount of hazardous chemical present at the facility at any time during the reporting period was greater than 10 times the reporting threshold;

LEVEL B: Amount of hazardous chemical present at the facility at any time during the reporting period was greater than 5, but less than or equal to 10 times the reporting threshold;

LEVEL C: Amount of hazardous chemical present at the facility at any time during the reporting period was greater than 1, but less than or equal to 5 times the reporting threshold.

Under §312, if one or more hazardous chemicals are present above thresholds at any time during the previous calendar year, an owner or operator of the facility is required to submit an Emergency and Hazardous Chemical Inventory Form, which may either be aggregate information by hazard category (Tier I) or specific information by chemical (Tier II). The form must report all hazards by category and must include information on all hazardous chemicals present at the facility during the previous calendar year in amounts that meet or exceed thresholds. For §312, the gravity levels are:

LEVEL A: For nonreporting situations: The amount of any hazardous chemical not included in the report was greater than 10 times the reporting threshold;

For reports timely submitted: 10 or more hazardous chemicals, which were required to be included in the report, were not included in said report.

LEVEL B: For nonreporting situations: The amount of any hazardous chemical not included in the report was greater than 5, but less than or equal to 10 times the reporting threshold;

For reports timely submitted: More than 5, but less than 10 hazardous chemicals, which were required to be included in the report, were not included in said report.

LEVEL C: For nonreporting situations: The amount of any hazardous chemical not included in the report was greater than 1, but less than or equal to 5 times the reporting threshold;

For reports timely submitted: 1 – 5 hazardous chemicals, which were required to be included in the report, were not included in said report.

Level C shall also apply to those submissions in which respondent's submitted form addresses all hazard categories necessary and all hazardous chemicals present above thresholds during the previous calendar year, but otherwise fails to meet the standard required by the Statute or Rule.

D. Circumstances

Circumstances refers to the potential consequences of the violation. The main objectives of the emergency notification provisions are to alert local, State, and Federal officials in the event of chemical accidents so that an appropriate emergency response action can be taken and to prevent injuries or deaths to emergency responders from exposure to chemicals. The main objectives of the emergency planning and community right-to-know provisions are to assist local and State committees in planning for emergencies and to make information on chemical presence and hazards available to the public. Thus, a respondent's failure to report in a manner that meets the standard required by the Statute or rule could result in a situation where there is potential for harm to human health and the environment. The potential for harm may be measured by:

- o the potential for emergency personnel, the community and/or the environment to be exposed to hazards posed by noncompliance, or

- o the adverse effect noncompliance has on the statutory or regulatory purposes or procedures for implementing the CERCLA §103/EPCRA program.

There are some requirements of the EPCRA or CERCLA programs which, if violated, may not be likely to give rise directly or immediately to a significant risk of exposure to hazards. Nonetheless, these requirements are fundamental to the continued integrity of the CERCLA and EPCRA programs. Violations of such requirements may have serious implications and merit substantial penalties where the violations undermine the statutory or regulatory purposes or procedures for implementing the EPCRA or CERCLA programs. Also, failure to provide the required information denies citizens their right to information regarding the chemical hazards that are present in the community.

After the extent and gravity of the violation have been determined (placing the proposed penalty in a given cell on the matrix), the circumstance factor is used to arrive at a specific penalty within the range for that cell. To incorporate the circumstances of the violation into the base penalty selection process, the case development team may choose any amount between, or including, one of the two end points for that cell. For example, a violation of §312 has been determined to have a Level 1 extent and a Level B gravity placing the proposed penalty in the matrix cell that contains the range of $16,500 - $13,200. The circumstances of the violation indicate that the potential for emergency personnel and the surrounding community to be at risk of exposure in the event of a release was high (the emergency personnel did not know of a chemical's presence and could not plan for the safety of the surrounding community in the event of a release). The case development team decides that the maximum amount for that cell is the appropriate base penalty.

The selection of the exact penalty amount within each range is left to the discretion of the enforcement personnel in any given case. In determining the circumstance level, consideration may be given to the relative proximity of the surrounding population, to the effect the noncompliance has on the LEPC's ability to plan for chemical emergencies, and any actual problems that first responders and emergency managers encountered because of the failure to notify (or submit reports) in a timely manner.

Table I. Base Penalty Matrices

CERCLA §103 and EPCRA §§302, 303, 304 and 312

Extent	Gravity Level A	Gravity Level B	Gravity Level C
Level 1:	$25,000 20,000	$16,500 13,200	$8,250 6,600
Level 2:	16,500 13,200	6,250 5,000	4,500 3,600
Level 3:	8,250 6,600	4,500 3,600	2,500 2,000

EPCRA §311 Violations

Extent	Gravity Level A	Gravity Level B	Gravity Level C
Level 1:	$10,000 8,000	$6,600 5,280	$3,300 2,640
Level 2:	6,600 5,280	2,500 2,000	1,800 1,440
Level 3:	3,300 2,640	1,800 1,440	1,000 800

V. ASSESSMENT OF MULTI-DAY PENALTIES

EPCRA §325 and CERCLA §109 authorize the Agency to assess penalties for violations on a per day basis. Two primary goals exist for using per day assessments: added deterrence and the need to receive the information sought. Use of a per day assessment may promote an expeditious return to compliance by creating disincentives for continued noncompliance and may be appropriate deterrence for those with a history of violations.

A number of situations may arise that would warrant the consideration of per day assessment of penalties. A violation may be so egregious that the case development team feels that a single day assessment will not be adequate. Situations where there is a continuing harm may also be cause for assessing penalties on a per day basis. These situations may warrant the assessment of the full base penalty (as calculated from the matrix) for each and every day the violative condition exists. Understanding that every case has its own peculiarities, the use of per day penalties will be at the discretion of the case development team. However, as with any other assessment, the justification for using, or not using, per day penalties should be incorporated into a memorandum to the case file.

Per day assessments can also be used in a more routine fashion. As was stated previously, one reason to use a per day assessment is to create incentives for violators to return to compliance as expeditiously as possible. One method to promote the expeditious return to compliance is to assess the base penalty for a single day and an additional smaller penalty from the date of the violation until the date of the compliance. Therefore, when a complaint is issued for a violation of §304(c), §311, or §312 and the situation warrants it, the complaint may seek a penalty based on calculations from the matrix and seek a per day assessment of a smaller penalty (e.g., $400 per day for each day the CERCLA §103 or EPCRA §304(a) notification, §304(c) report, §311 MSDS, or §312 inventory form continues) from the date of the violation until the required reports are submitted. The case development team should require the respondent to send EPA copies of required submissions to verify compliance. This approach normally should be used for first time violators.

For second and subsequent violations, CERCLA §109 and EPCRA §325 authorize the Agency to assess penalties of up to $75,000 per day for each and every day violations of CERCLA §103 and EPCRA §§304(a) and 304(c) continue. Per day penalties may be calculated by trebling the amount of the base penalty calculated in the matrix and assessing that amount each day the violation continues.

Section 325 of EPCRA does not authorize a special category of penalties for second and subsequent violations of §§311 and 312. Using the per day assessment of penalties should be adequate to handle second and subsequent violations of §§311 and 312. The per day assessment for a second or subsequent violation should run from the date the violation began until the date the violative condition ends. For second time violations, the base penalty should be assessed for the first day of violation and 50 per cent of the calculated base penalty should be assessed for every other day the violation continues. Third and subsequent violations should be assessed the full statutory daily amount (See also the Section on Prior History of Violations).

VI. CALCULATION OF PENALTY FACTORS RELATING TO THE VIOLATOR

The base penalty reflects the overall seriousness of the violation. The reasons the violation was committed, the intent of the violator, and other factors related to the violator are not considered in choosing the appropriate penalty from the matrix. However, any system for calculating penalties should have enough flexibility to make adjustments for legitimate differences between similar violations. CERCLA §109 and EPCRA §325 require (for Class I violations of CERCLA §103 and EPCRA §304) the Agency to consider certain factors related to the violator. Specifically, in calculating a penalty the Agency must consider ability to pay/continue in business, any prior history of such violations, the degree of culpability, economic benefit or savings (if any), and such other matters as justice may require (See Footnote 1). These factors, while not exculpatory, need to be considered in every penalty assessment.

With respect to settlement, before EPA considers adjusting the penalty contained in the complaint and applies the factors relating to the violator, it may be necessary, under certain circumstances, for enforcement personnel to recalculate the base penalty. If new information becomes available after the issuance of the complaint that makes it clear that the initial calculation of the penalty contained in the complaint is in error, enforcement personnel should adjust this figure (either up or down). The basis for any recalculation of the base penalty made at this time should be documented on the Penalty Calculation Worksheet. For example, if after the issuance of the complaint, information is presented that indicates that much less of the chemical is involved than was believed when the complaint was issued, it may be appropriate to recalculate the base penalty.

In applying the factors relating to the violator, it must be kept in mind that the statutory maximums of $25,000 per violation (§304 Class I penalty), $25,000 per violation per day (§304 Class II penalty and §312) or $10,000 per violation per day (for §311) cannot be exceeded for any violation no matter which adjustment factors apply.

A. Ability To Pay/Continue In Business (Downward Adjustment Only)

The Agency will generally not request penalties that are clearly beyond the financial means of the violator. However, EPA reserves the option, in appropriate circumstances, of seeking a penalty that might put a company out of business[2]. For example, even when there is an ability-to-pay problem, it is unlikely that EPA would reduce a penalty when a facility refuses to correct a serious violation or where a facility has a long history of violations. That long history would demonstrate that less severe measures are ineffective.

As mentioned previously, the penalty amounts reflected in the matrix assume that the violator has the ability to pay. The financial ability adjustment will normally require that the Agency receive a significant amount of information specific to the violator. The case development team should assess this factor after commencement of the negotiation with the violator as more information becomes available. The burden to demonstrate inability to pay, as with the burden to demonstrate any other mitigating factor, rests with the violator. If the violator fails to provide sufficient information, then the case development team should continue to assume ability to pay exists.

There are several sources available to assist the Regions in determining a firm's ability to pay. The National Enforcement Investigations Center (NEIC) can help obtain information assessing the financial ability to pay of publicly held corporations. Additionally, enforcement personnel should acquaint themselves with the Office of Enforcement's ABEL, the Agency's computer model that helps analyze ability to pay for compliance, clean-up, and/or penalties. Although ABEL was designed with privately held corporations in mind, it will soon be expanded to include other forms of business entities and it may serve as an adjunct to other programs available through NEIC (e.g., the Superfund Financial Assessment System).

If an alleged violator raises the ability to pay argument as a defense in its answer, or in the course of settlement negotiations, it shall present sufficient documentation to permit the Agency to establish such inability. Appropriate documents will include the following, as the Agency may request, and will be presented in the form used by the respondent in its ordinary course of business:

- Tax returns
- Balance sheets
- Income statements
- Statements of changes in financial position
- Statements of operations

---

[2] Ability to continue in business must be considered, as a matter of law, only when assessing penalties for violations of EPCRA §304 under EPCRA §325(b)(2).

- Retained earnings statements
- Loan applications, financing agreements, security agreements
- Annual and quarterly reports to shareholders and the SEC, including 10 K reports

Such records are to be provided to the Agency at the respondent's expense and must conform to generally recognized accounting procedures. The Agency reserves the right to request, obtain, and review all underlying and supporting financial documents that form the basis of these records to verify their accuracy. If the alleged violator fails to provide the necessary information, and the information is not readily available through other sources, then the violator will be presumed to have the ability to pay.

B. Prior History of Violations (Upward Adjustment Only)

The Base Penalty Matrix is designed to apply to first time offenders. Where a violator has a history of similar violations under CERCLA and EPCRA at the same or a different site, this is usually clear evidence that the previous penalty did not provide sufficient deterrence. For purposes of this policy, the Agency interprets "prior violations" to mean violations of CERCLA (for releases) or EPCRA only. The following rules apply to evaluating the history of prior violations:

o A prior violation is considered to be any act or omission for which a formal enforcement response has occurred regardless of whether or not respondent admits to the violation (e.g complaint, default judgment, consent decree, or consent agreement/final order).

o To be considered a prior violation, the final order, default judgment, or consent decree must have been entered within five (5) years of the present violation.

o In the case of large corporations with many divisions or wholly-owned subsidiaries, it may be difficult to determine whether a previous instance of noncompliance should trigger upward adjustments to the base penalty. New ownership often raises similar problems. In general, enforcement personnel should begin with the assumption that if the same corporation was involved, adjustments for history of noncompliance should apply. The Agency may find a consistent pattern of noncompliance by many divisions or subsidiaries of a corporation even though the facilities are at different geographic locations. This often reflects, at best, a corporate-wide indifference to environmental compliance. Consequently, the adjustment for history of noncompliance should apply unless the violator can demonstrate that the other violating facilities are independent. In the case of wholly- or partially-owned subsidiaries, the violation history of the parent corporation shall apply to its subsidiaries and that of the subsidiaries to the parent.

For the purposes of this penalty policy, a violation of §313 will count as a prior violation if the §313 violation occurred in one of the previous five years. The situation may arise where a §313 enforcement action will lead to other EPCRA enforcement actions being filed against the same facility arising from the same set of facts. If the owner or operator entered into a consent agreement with EPA for the §313 violation and in that consent agreement certified their compliance with all of EPCRA requirements and later they were found to be in violation of §§302-312, the EPCRA §313 violation may be counted as a prior violation. Also, if they falsely certify their compliance, the respondent could be criminally liable. If this situation arises, contact the regional office that handles criminal investigations.

As noted in the section on multi-day assessments, for second or subsequent violations of CERCLA §103 and EPCRA §304, a penalty of up to $75,000 per violation per day is authorized. For second or subsequent violations of these requirements, treble the amount shown at the appropriate position in the base penalty matrix. If the prior violation was for a non-§304 EPCRA requirement, the case development team should consider assessing the base penalty for each day of violation. For second time violations, the base penalty should be assessed for the first day of the violation and 50 per cent of the calculated base penalty should be assessed for every other day the violation continues. Third and subsequent violations should be assessed the full statutory daily amount (See also the section on multi-day penalties).

C. <u>Degree of Culpability</u> (Upward or Downward Adjustment)

The existence of a violation is established without a showing of failure to adhere to a standard of care. As with other statutes, EPA pursues a policy of strict liability in penalizing for a violation. Nonetheless, under the penalty system in this policy, the base penalty may be increased, decreased or remain the same depending on the violator's culpability.

Two concepts that underlie culpability are the violator's knowledge of the requirement and the violator's control over the violative act. The lack of knowledge of a particular requirement would not necessarily reduce culpability. To do so would encourage ignorance of the law. The test under CERCLA §103 and EPCRA §§304, 311, and 312 will be whether the violator knew or should have known of the CERCLA/EPCRA requirements or that the general nature of his operation deals with hazardous chemicals. A reduction in penalty based upon lack of knowledge may only occur where a reasonably prudent and responsible person in the violator's position would not have known that the conduct was violative of CERCLA or EPCRA.

The amount of control that the violator had over how quickly the violation was remedied is relevant in certain instances. Specifically, if correction of the violative condition was delayed by circumstance that the violator can clearly show were not reasonably foreseeable and out of its control, the penalty may be reduced.

The violator can manifest good faith by promptly identifying and reporting noncompliance <u>before the Agency detects the violation.</u> This situation may justify mitigation of a penalty. Lack of good faith, on the other hand, can result in an increased penalty. No downward adjustment should be made for the respondent's efforts to comply after the Agency has detected a violation. Indeed, failure to take such actions may justify upward adjustment of the penalty.

If a respondent relies on written guidance by the state or EPA that an activity will satisfy EPCRA or CERCLA §103 requirements and later it is determined that the activity does not comply with EPCRA or CERCLA, a downward adjustment in the penalty may be warranted, but only if the respondent can substantiate its claim that it relied on those assurances in good faith. On the other hand, claims by a respondent that "it was not told" by EPA or the State that it was out of compliance should not justify any downward adjustment of the penalty.

Any prior contact that EPA, the State or LEPC has had with the respondent including, but not limited to, documented phone contacts, Administrative Orders under EPCRA §§302 and 303, Notices of Violation, warning letters, contact under EPCRA §313, and/or respondent's attendance at EPCRA seminars may be used to help determine the culpability of the respondent. Formal enforcement actions against the respondent that result in issuance of a consent decree, final order or default order should be counted under the Prior History of Violations determination.

For purposes of CERCLA §103 and EPCRA §§304, 311 and 312, three levels of culpability have been assigned:

Level 1: The violator had prior knowledge of EPCRA and its reporting requirements as evidenced by its attendance at an EPCRA seminar or workshop, or having been previously contacted by EPA, the SERC, or LEPC through a documented phone conversation concerning EPCRA, an EPCRA informational letter, EPCRA warning letter, EPCRA §313 activities, EPCRA Notice of Violation, etc. -- Increase the base penalty up to 25%.

Level 2: The violator did not comply either due to lack of knowledge of the requirement, lack of management requirements in systems, or failure to adhere to internal procedures. -- No adjustment to the base penalty.

Level 3: The violator attempted to comply properly or self-confessed before the Agency detected the violation. -- Decrease the base penalty up to 25%.

It is anticipated that most cases will present Level 2 culpability. However, if it can be shown that the facility had previous knowledge of EPCRA or had previously participated in an EPCRA training or seminar or received any outreach literature, notification, warning letter, etc., from EPA, the SERC, or LEPC regarding EPCRA reporting requirements, a Level 1 culpability ranking may be considered.

D. Economic Benefit or Savings

EPA should consider any economic benefit from noncompliance that accrues to the violator when assessing penalties. Whenever there is an economic incentive to violate the law, it encourages noncompliance and thus weakens EPA's ability to implement the Acts and protect human health and the environment. The violator should not benefit from its violative acts. An economic benefit component should be calculated and added to the base penalty (but not to exceed the statutory maximum) when a violation results in any economic benefit to the violator. However, the base penalty cannot exceed the statutory maximum.

For EPCRA §§304(c), 311, and 312 reporting violations, the economic benefit or savings typically is derived from the estimated cost of producing and submitting the reports. The economic benefit derived from failure to provide emergency notification (e.g., the cost of a phone call) is considered negligible. The economic benefits for failure to submit §§311 and 312 reports include the costs of producing the reports and any filing fees that are imposed by States.

The Regulatory Impact Analysis (RIA) for the §§311/312 regulation establishes unit costs for producing the required reports (see Table II, Page 30, *Infra*). These cost estimates should be used unless more accurate data is available. Costs are disaggregated into costs associated with rule familiarization, establishment of filing systems, threshold effects, preparation and submission of required reports. In using this information to determine economic savings for multiple violations, the variable costs should be counted once only and the fixed costs counted for each chemical violation.

It is anticipated that most of the savings associated with these violations in a number of cases may be negligible. In the interest of simplifying and expediting the enforcement action, enforcement personnel may forego calculating the economic benefit if it appears to be less than $2,500. However, this decision should be documented in the narrative penalty justification kept in the case file. If it looks to be close to, or above $2,500, the economic benefit should be calculated using BEN. If the BEN evaluation derives an economic benefit above $2,500, that amount should be included in the penalty.

It is generally the Agency's policy not to settle cases for an amount that is less than the economic benefit of non-compliance. However, this civil penalty policy sets out four general areas where settling the case for less than the economic benefit may be appropriate. These include situations when:

- there are compelling public concerns that would not be served by taking a case to trial;
- it is highly unlikely, based on the factors of the case as a whole, that EPA will be able to recover the amount of the economic benefit in litigation;
- the company has documented an inability to pay more than the amount of the estimated economic benefit; or
- the economic benefit is insignificant (i.e., < $2,500).

Table II. Cost Associated with EPCRA §311 and §312 Reports

COSTS ASSOCIATED WITH §§311 AND 312 REPORTS

Fixed Unit Costs under §311

Copy, handle and mail MSDS	$1.84
File MSDS	$2.84
MSDS Cover Letter	$14.56
	$18.73

Fixed Unit Costs under §312

Decision on Tier I/II	$239.25
Hazard Classification	$7.48
Typing and QA/QC	$67.81
Preparing forms	$6.18
Copying & Mailing	$5.37
	$326.09

Variable Unit Costs for Manufacturers*

Employees	0 – 19	20 – 99	100 - 249	>250
§311				
Rule Familiarization	$43.50	$65.25	$99.78	$146.81
Filing System	$400.88	$601.32	$901.98	$1352.97
Threshold Effects	$27.20	$40.80	$61.19	$91.79
§312				
Rule Familiarization	$43.50	$65.25	$99.78	$146.81

* Unit costs in the non-manufacturing sector for rule familiarization, filing system, and threshold effects for all facility size categories are assumed to be comparable to the unit costs in the manufacturing sector for the same activities in the size category of 0 to 19 employees.

E. Other Matters as Justice May Require

This policy acknowledge that no two cases are exactly alike. Unique circumstances above and beyond those taken into account by the factors discussed in the previous sections may be significant in determining the appropriateness of a penalty. The following discussions address some circumstances that may affect the settlement penalty amount.

i). Delisting Reductions

If the Agency proposes the delisting of a chemical on the extremely hazardous substance (EHS) list by a Federal Register Notice, the Agency may settle cases involving the proposed delisted chemical under terms which provide for a 25% deferral of the initial penalty calculated for any EPCRA §§302, 303, 304, 311, or 312 violation involving that chemical. Note, that if the chemical does become delisted, reporting obligations under §§311 and 312 may still apply, however, the applicable threshold would be the 10,000 pound threshold which normally applies to other "hazardous chemicals" under the OSHA Hazard Communication Standard. The deferral policy is only applicable to chemicals proposed for delisting before or during the pendency of the enforcement action. The penalty deferred becomes due and owing 30 calendar days after publication of the Agency's decision to retain the chemical on the extremely hazardous substance list. If the Agency's final published decision is to delist the chemical, the deferral becomes a reduction in penalty which is in addition to any other possible reductions possible in this policy.

ii). Environmentally Beneficial Expenditures

Instances may arise where a violator will offer to make expenditures for environmentally beneficial projects above and beyond those required by law. In these instances, it may be appropriate to accept a lower penalty amount for settlement in light of the totality of the agreement. The Agency, in settling penalty action in the U.S. District Courts under the Clean Air and Water Acts, has determined that considering such expenditures is consistent with the purpose of civil penalty assessment in certain cases. The same rationale applies to penalties that are assessed in administrative settlements. In the past, the Agency has used its enforcement discretion to mitigate proposed penalties for some environmentally beneficial projects proposed and implemented by the respondent. In applying this penalty policy, this mitigation is completely discretionary.

This adjustment constitutes a basis for accepting a lower cash penalty amount. Before any proposed adjustments are incorporated into a settlement, the case development team should ensure that all of the following conditions are met:

- No adjustments can be given for activities that currently are or will be required under the current law or are likely to be required under existing statutory authority in the foreseeable future (e.g., through rulemaking).

- The majority of the project's environmental benefit should accrue to the general public rather than to the source.

- The project cannot be something that the violator could reasonably be expected to do as part of sound business practices.

- EPA must not lower the amount it decides to accept in penalties by more than the after tax net-present value of the project. (The after tax net-present value of a project can be calculated on BEN.)

- The project proposed by the Respondent should promote the goals of EPCRA: to increase emergency planning, preparedness, and response or to increase public awareness of EPCRA.

- Environmentally beneficial expenditures may include those expenditures that go to a SERC or LEPC for a designated use to further the goals of EPCRA.

- The mitigation for environmentally beneficial expenditures may not reduce the penalty below the economic benefit of noncompliance.

In all cases where alternative payments are accepted, the case development team should document that each of the conditions mentioned above are met and include this documentation in the case file. Additionally, the case development team should take into account the following:

- The project should not require a large amount of EPA oversight;

- The project should receive stronger consideration if it takes place in the locality in which the facility is located;

- The company should agree that any publicity it disseminates regarding its funding of the project must include a statement that such funding is in settlement of a lawsuit brought by EPA.

Each alternative payment plan must entail an identified project to be completely performed by the defendant. Under the plan, EPA must not hold any funds that are to be spent at the Agency's discretion. The final order, decree or judgment should state what financial penalty the violator is actually paying and describe, as precisely as possible, the environmentally beneficial project the violator is expected to perform.

iii). Settlement Considerations

Any reductions in penalties are to be made in accordance with this penalty policy. In settling cases, if the case development team wishes to enter into an agreement with the company to an audit of the company's facility(ies), the consent agreement and consent order should contain related provisions. Any additional violations identified during the audit may be assessed penalties in accordance with this penalty policy and may include stipulated penalties. However, reductions for voluntary disclosure may be made as appropriate.

iv). Documentation

Any mitigation of the proposed penalty must be documented in the case file. A narrative justification and a revised penalty calculation worksheet should document the amount of the penalty mitigated and the justifications for the mitigation based on the statutory factors. A penalty calculation worksheet and a narrative explanation worksheet are included in Appendix I.

VII. APPENDIX I

PENALTY CALCULATION WORKSHEET

Respondent:______________________ Complaint DCN: __________
Count #:______________________ Inspection Date: __________
Chemical Name/CAS:______________________ RQ/TPQ: __________
Violation:______________________

NATURE: a) Emergency Response (CERCLA §103/EPCRA §304) __________
b) Planning/Right-to-know (§§302, 303, 311, 312) __________

EXTENT: Time passed from deadline to performance of required action in hours or days, specify: __________ or the amount of deviation from the requirement. Matrix level __________

GRAVITY: 1) Amount of chemical involved in violation (lbs.) __________
2) Divide amount in 1) by __________ [RQ/TPQ/Threshold (circle one)] = __________ . Matrix level __________

CIRCUMSTANCES: 1) Likelihood of exposure to hazards posed by violation, or
2) Adverse effect violation has on implementing the EPCRA program: High___ Medium___ Low___. Specify choice of penalty amount from range listed for the cell of the matrix ________.

1. Base Penalty ....$ __________
2. Culpability (% increase or decrease +/- ....). $
3. Prior History: §§304/103: treble bass amount per day penalty
§§311/312: per day penalty
4. If per day, multiply line 1 by days of noncompliance ...$ __________
If treble, multiply Line 1 by 3 ....$ __________
5. Add lines 2 and 4 ....$ __________
6. Economic gains from noncompliance ....$ __________
7. Add lines 5 and 6 ....$ __________
8. Other adjustments as justice may require ....$ __________
9. Total penalty* (line 7 +/- line 8) ....$ __________

*For first time violators, total penalty cannot exceed $25,000 per violation per day or $10,000 per violation per day (for §311).

Repeat procedure for each violation.

Prepared by:______________________ Signature:______________________
Date: ______________________ Page ______ of ______

NARRATIVE EXPLANATION

[SECTION VIOLATED] Date:________________

A. FACTORS THAT APPLY TO THE VIOLATION

NATURE:

EXTENT:

GRAVITY:

CIRCUMSTANCES:

B. FACTORS THAT APPLY TO THE VIOLATOR

ABILITY TO PAY:

PRIOR HISTORY OF VIOLATIONS:

DEGREE OF CULPABILITY:

ECONOMIC BENEFIT OR SAVINGS:

OTHER MATTERS AS JUSTICE MAY REQUIRE:

List of EPA Enforcement Actions for Violations of Sections 304-312

ADMINISTRATIVE ENFORCEMENT ACTIONS
CERCLA §103 AND EPCRA §§304-312
FY89-92
(as of September 1, 1992)

REG	DATE ISSUED	NAME	EPCRA Violations	CERCLA Violations	PROPOSED EPCRA $	PROPOSED CERCLA $	FINAL EPCRA $	FINAL CERCLA $	DATE
1	09/30/88	ALL REGIONS	304(a),(c)	103(a)	97000	25000	69840	20000	12/01/89[1]
3	12/07/88	MURRY'S INC.	304(a),311,312	**	25000	0	21250	0	05/25/89
3	12/07/88	MURRY'S INC.	**	103(a)	0	26000	0	15550	05/25/89
4	04/27/89	CONS. MIN. INC	304(a),311,312	**	30000	0	30000	0	06/27/89
5	05/15/89	BF GOODRICH	304(a),(c)	103(a)	6000	3000	6000	3000	08/09/89
5	05/15/89	KRAFT FOODS	304(a),(c)	103(a)	16000	8000	8000	8000	12/20/89
5	05/15/89	COMM. EDISON	304(a),(c)	103(a)	10000	5000	6666	3334	09/16/89
8	06/30/89	TRISTATE MINT	304(a),(c)	103(a)	50000	25000	0	0	XX/XX/XX
1	08/04/89	CATELLI FOODS	304(a),311,312	103(a)	36000	4000	19950	3000	05/15/90
1	09/27/89	CHAMPION INTRL	304(a)	103(a)	10000	10000	6000	6000	10/31/90
1	09/18/89	CARBOLABS INC.	302-3,311,312	**	36000	0	20600	0	01/03/90
2	09/21/89	BP OIL CO.	304(a)	103(a)	175000	175000	51000	51000	04/30/90
1	12/15/89	SIMS INC.	304,311,312	103(a)	33200	15000	3200	0	10/31/90
4	03/15/90	BREWER GOLD	**	103(a)	0	25000	0	25000	03/15/90
6	04/17/90	EUTECTICS INC.	304,311,312	**	31350	0	30000	0	10/26/90
6	05/10/90	BURLINGTON NORTHERN	**	103(a)	0	25000	0	17000	11/27/90
3	05/04/90	GOLD CREST	304,311,312	103(a)	13109	59800	0	0	XX/XX/XX
1	06/25/90	GRANITE STATE	304,311,312	103(a)	106200	25000	0	0	XX/XX/XX
1	06/26/90	LACROIX FOOD	304,311,312	103(a)	33680	6600	5998	4950	04/24/91
1	06/26/90	PEASE & CURREN	311,312	**	43720	0	0	0	XX/XX/XX
2	06/26/90	ANAQUEST INC.	304	103(a)	50000	25000	0	0	XX/XX/XX
2	06/26/90	O-CEL-O/GEN MILLS	304	103(a)	119700	37350	82250	30250	10/18/90
3	06/26/90	MAP INTERNATIONAL	304	103(a)	80000	20000	0	0	XX/XX/XX
3	06/26/90	J.O. TANKERS	**	103(b)	0	20000	0	0	XX/XX/XX
3	06/26/90	VALLEY BUMPERS	**	103(a)	0	20000	0	0	XX/XX/XX
4	06/27/90	CITRUS HILL MANUF	304	103(a)	25000	15000	0	0	XX/XX/XX
4	06/27/90	MEADOW GOLD DAIRIES	304	103(a)	198000	49500	0	0	XX/XX/XX
4	06/27/90	HERCULES, INC.	**	103(a)	0	15000	0	0	XX/XX/XX
5	06/27/90	DOVE INTERNATIONAL	304	103(a)	91500	25000	50000	25000	09/19/90
5	06/27/90	DIVERSE PLASTICS	311,312	**	34650	0	34650	0	10/12/90
5	06/27/90	GENERAL ELECTRIC	304	103(a)	75000	25000	65000	25000	07/27/90
5	06/27/90	BERKSHIRE FOODS	304,311,312	103(a)	156100	25000	31900	2900	01/24/91
5	06/27/90	MEAD FINE PAPER	304	103(a)	61875	20625	0	0	XX/XX/XX
6	06/25/90	ALFORD REFRIGERATION	304,311,312	103(a)	66000	0	40000	0	09/05/91
6	06/25/90	GYRO CHEMICAL	304,311,312	103(a)	84350	16500	32660	6220	09/05/91
6	06/25/90	BASF CORPORATION	304	103(a)	43250	8250	52000	0	03/21/91
7	06/27/90	DOE RUN COMPANY	304	103(a)	40500	22500	40500	22500	02/07/91
8	06/27/90	THORO PRODUCTS	304,311,312	103(a)	68000	16500	31500	3500	05/19/92
9	06/27/90	NOGALES ICE CO	304,311,312	103(a)	15840	6600	0	0	XX/XX/XX
10	06/27/90	LIQUID CARBONIC	304,311,312	103(a)	60520	23000	0	0	XX/XX/XX
1	09/28/90	WYMAN-GORDON CO	312, 313	103(a)	472000	6600	0	0	XX/XX/XX
1	09/28/90	STOP & SHOP CO	304,311,312	103(a)	48840	13200	0	0	XX/XX/XX
1	09/28/90	WEST LYNN CREAMERY	302,303,312	**	60000	0	0	0	XX/XX/XX
SUBTOTAL PAGE 1 $:					2,603,384	848,025	738,964	272,204	

1 U.S. Court of Appeals, 1st circuit, upheld EPA imposed penalty for the CERCLA count.

REG	DATE ISSUED	NAME	EPCRA Violations	CERCLA Violations	PROPOSED EPCRA $	PROPOSED CERCLA $	FINAL EPCRA $	FINAL CERCLA $	DATE
5	10/05/90	LOMAC, INC.	304	103(a)	82500	41250	56400	37600	03/26/91
5	10/19/90	MINERAL MET INC.	311,312	**	387750	0	130,000	0	04/24/92
2	11/23/90	SCHENECTADY CHEMICAL	304	103(a)	45000	10000	36250	10250	05/17/91
1	12/28/90	DELTA SURPRENANT WIRE	311,312	**	123520	0	0	0	XX/XX/XX
1	12/28/90	BRISTOL CO WATER AUTH.	304,312	103(a)	26400	6600	7000	0	05/14/91
5	01/10/91	NATIONAL STEEL CORP	304	103(a)	75000	25000	0	0	XX/XX/XX
2	01/17/91	CALGON CORPORATION	304	103(a)	83000	41500	0	0	XX/XX/XX
2	01/17/91	ICI AMERICAS, INC	304	103(a)	75000	50000	0	0	XX/XX/XX
2	01/28/91	MERCK & COMPANY, INC	304	103(a)	16500	8250	0	0	XX/XX/XX
6	02/01/91	CENTRAL PROPANE, INC	311,312	**	29700	0	8910	0	05/14/91
6	02/01/91	BLACK GOLD ACIDIZING	311,312	**	14850	0	8910	0	04/12/91
8	02/05/91	CHEMICAL SALES COMPANY	**	103(a)	0	7425	0	7425[2]	03/17/92
1	03/29/91	TORRINGTON COMPANY	311,312	**	50520	0	0	0	XX/XX/XX
5	04/15/91	ZENITH ELECTRONICS CORP	304	103(a)	66000	16500	55600	13900	08/01/91
3	04/02/91	KORMAN, HYMAN INC	311,312	**	105000	0	0	0	XX/XX/XX
3	04/02/91	GENICOM CORPORATION	**	103(a)	0	49750	0	0	XX/XX/XX
3	04/02/91	GENICOM CORPORATION	304	**	99500	0	0	0	XX/XX/XX
3	04/04/91	HOLLY FARMS FOODS, INC	**	103(a)	0	20000	0	0	XX/XX/XX
3	04/04/91	HOLLY FARMS FOODS, INC	304	**	40000	0	0	0	XX/XX/XX
6	04/23/91	BUCHANON, EDGE & HUGHES	311,312	**	23150	0	4000	0	09/05/91
2	05/13/91	MOBIL OIL CORP	304	103(a)	100000	50000	0	0	XX/XX/XX
2	05/13/91	MOBIL OIL CORP	304	**	100000	0	0	0	XX/XX/XX
2	05/13/91	MOBIL OIL CORP	304	**	325000	0	0	0	XX/XX/XX
8	06/14/91	KOCH HYDROCARBON	304	103(a)	100000	25000	22500	70275	11/07/91
5	08/01/91	RHONE-POULENC CHEM CO	**	103(a)	0	29700	0	0	XX/XX/XX
9	07/15/91	PIONEER CHLOR ALKALI CO,INC	304	103(a)	169200	25000	80808	25000	04/24/92
6	09/04/91	LYTLE FERTILIZER CO.	**	103(a)	0	15000	0	0	05/11/92
6	09/04/91	ANALYTICAL SYSTEMS CO.	312	**	20000	0	0	0	02/13/92
6	09/05/91	OKLAHOMA STEEL & WIRE	312	**	16500	0	0	0	02/24/92
8	09/23/91	BRUSH WELLMAN, INC.	304	103(a)	90000	42500	35600	8900	04/13/92
5	09/30/91	INDUSTRIAL SCRAP CO.	304	103(a)	33000	8250	0	0	XX/XX/XX
5	10/23/91	KEIL CHEMICAL CO.	304	**	39250	0	0	0	XX/XX/XX
6	11/14/91	TREAT-RITE LABS, INC.	312	**	25000	0	3000	0	03/09/92
5	11/19/91	BORDEN CHEM & PLASTICS CO.	304	103(a)	800000	200000	0	0	XX/XX/XX
5	11/19/91	GRIMMER CONSTRUCTION CO.	304	103(a)	93720	16500	0	0	XX/XX/XX
8	12/19/91	KENNECOTT UTAH COPPER CORP.[3]	304	103(a)	269850	22500	0	0	XX/XX/XX
6	12/27/91	FILGO OIL CO.	312	**	2500	0	0	0	06/05/92
6	12/27/91	JACK RAY & SONS OIL CO. INC.	312	**	8250	0	4800	0	02/24/92
8	01/27/92	SEVEN PEAKS RESORT	304,311,312	103(a)	61,740	14,850	0	0	XX/XX/XX
	&								
SUBTOTAL PAGE 2 $					3,597,400	725,575	453,770	173,350	

[2] Respondent filed for bankruptcy protection under Chapter 11 of the Bankruptcy Code on February 21, 1992.

[3] Complaint also alleges TSCA violations and seeks penalties of $1,129,000 under TSCA.

ADMINISTRATIVE ENFORCEMENT ACTIONS
CERCLA §103 AND EPCRA §§304-312
FY89-92
(as of September 1, 1992)

REG	DATE ISSUED	NAME	EPCRA Violations	CERCLA Violations	PROPOSED EPCRA $	PROPOSED CERCLA $	FINAL EPCRA $	FINAL CERCLA $	DATE
6	01/31/92	CHEMCOM INC. OF THE SOUTH	**	103(a)	0	8250	0	1000	05/11/92
6	01/31/92	DORSETT BRO. CONCRETE INC.	312	**	8250	0	1500	0	02/24/92
6	02/06/92	NATIONAL FREIGHT INC.	**	103(a)	0	8250	0	2900	03/30/92
6	02/07/92	LYONDELL PETROCHEMICAL, INC.	**	103(a)	0	25000	0	1000	05/11/92
6	02/07/92	DOW CHEMICAL CO., INC.	**	103(a)	0	8250	0	3700	08/27/92
6	03/04/92	HOECHST CELANESE, INC.	**	103(a)	0	8250	0	0	XX/XX/XX
6	03/09/92	EXXON CHEMICAL CO., INC.	**	103(a)	0	25000	0	2500	07/13/92
6	03/12/92	CHEVRON CHEMICAL CO., INC.	**	103(a)	0	25000	0	17500	08/21/92
6	03/13/92	BLUEBONNET WASTE CONTROL, INC	312	**	8250	0	3000	0	06/01/92
6	03/13/92	GENERAL DYNAMICS CORP.	304	103(a)	58000	58000	0	0	XX/XX/XX
6	04/16/92	SWIFT ECKRICH	304	103(a)	?	?	1750	1750	07/15/92
6	05/14/92	CETEC PROCESSING	312	**	8250	0	1500	0	06/30/92
6	05/16/92	ADOBE INDUSTRIES, INC.	312	**	8250	0	1500	0	07/13/92
6	05/16/92	EL PASO WIRE, INC.	312	**	8250	0	3000	0	07/15/92
6	05/16/92	CONSUMER ICE COMPANY	312	**	8250	0	0	0	XX/XX/XX
5	06/09/92	ELI LILLY & CO.	304	103	91500	25000	0	0	XX/XX/XX
10	06/11/92	QUIL CEDA TANNING CO.	302, 304	**	0	0	0	0	XX/XX/XX
6	06/12/92	EXCELEX CORPORATION	312	**	25000	0	3000	0	07/13/92
6	06/24/92	SID RICHARDSON CARBON & GASOLINE CO.-BORGER PLT.	**	103(a)	0	16500	0	3000	08/13/92
6	06/29/92	OCCIDENTAL CHEMICAL CORP.	**	103(a)	0	25000	0	0	XX/XX/XX
10	07/15/92	BP OIL COMPANY	304	**	138000	0	0	0	XX/XX/XX
10	07/15/92	EMERALD CITY CHEMICAL CO.	312	**	75000	0	0	0	XX/XX/XX
10	07/15/92	WEYERHAEUSER COMPANY	304	103(a)	50000	25000	0	0	XX/XX/XX
6	07/17/92	LADY LUCK OIL CO.	312	**	6600	0	1500	0	08/13/92
5	08/25/92	HAVILAND PRODUCTS CO.	304	103(a)	75000	25000	0	0	XX/XX/XX
5	08/26/92	GRAMMER INDUSTRIES, INC.	304	103(a)	100000	25000	0	0	XX/XX/XX
6	08/26/92	ELF ATOCHEM N.A., INC.	312	**	8250	0	0	0	XX/XX/XX
6	08/26/92	GULF COLD STORAGE CO.	312	**	8250	0	0	0	XX/XX/XX
6	08/26/92	LOOP COLD STORAGE CO.	312	**	8250	0	0	0	XX/XX/XX
6	08/26/92	MID VALLEY INDUSTRIES	312	**	8250	0	0	0	XX/XX/XX
6	08/26/92	POTTER PAINT CO. OF TX, INC.	312	**	8250	0	0	0	XX/XX/XX
6	08/26/92	SOUTH TEXAS CHLORINE	312	**	8250	0	0	0	XX/XX/XX
6	08/27/92	BOURG CHEMICAL	312	**	1000	0	0	0	XX/XX/XX
6	08/27/92	CENTURY ICE CO.	312	**	8250	0	0	0	XX/XX/XX

$

	PROPOSED EPCRA $	PROPOSED CERCLA $	FINAL EPCRA $	FINAL CERCLA $
SUBTOTAL PAGE 3 $	727,350	307,500	16,750	33,350
COLUMN TOTALS $	6,928,134	1,881,100	1,209,484	478,904
TOTAL $ ASSESSED AND COLLECTED	8,809,234		1,688,388	

EPA Enforcement Policy for Section 313

ENFORCEMENT RESPONSE POLICY

FOR SECTION 313 OF THE

EMERGENCY PLANNING AND COMMUNITY RIGHT-TO-KNOW ACT (1986)

And

SECTION 6607 OF THE

THE POLLUTION PREVENTION ACT (1990)

Issued by the

Office of Compliance Monitoring

of the

Office of Prevention, Pesticides and Toxic Substances

United States Environmental Protection Agency

August 10, 1992

TABLE OF CONTENTS

SUBJECT **PAGE**

INTRODUCTION

On December 2, 1988, the U.S. Environmental Protection Agency (EPA) issued an Enforcement Response Policy for addressing violations of Section 313 of the Emergency Planning and Community Right-to-Know Act. Since that time, EPA has identified opportunities for refining and adding clarity to that policy. This revised enforcement response policy incorporates three years of enforcement experience with Section 313 of the Emergency Planning and Community Right-to-Know Act.

This policy is immediately applicable and will be used to calculate penalties for all administrative actions concerning EPCRA Section 313 issued after the date of this policy, regardless of the date of the violation.

The Emergency Planning and Community Right-to-Know Act, (EPCRA), also known as Title III of the Superfund Amendments and Reauthorization Act of 1986, contains provisions for reporting both accidental and nonaccidental releases of certain toxic chemicals. Section 313' (§313) of EPCRA requires certain manufacturers, processors, and users of over 300 designated toxic chemicals to report annually on emissions of those chemicals to the air, water and land. The Pollution Prevention Act (PPA) of 1990 requires additional data and information to be included annually on Form R reports beginning in the 1991 reporting year, for reports which are due on July 1, 1992. These reports must be sent to the U.S. Environmental Protection Agency (EPA) and to designated state agencies. The first reporting year was 1987, and reports were due by July 1, 1988, and annually by July 1 thereafter. The U.S. EPA is responsible for carrying out and enforcing the requirements of §313 of EPCRA and the PPA and any rules promulgated pursuant to EPCRA and the PPA.

Section 325(c) of the law authorizes the Administrator of the EPA to assess civil administrative penalties for violations of §313. Any person (owner or operator of a facility, other than a government entity) who violates any requirement of §313 is liable for a civil administrative penalty in an amount not to exceed $25,000 for each violation. Each day a violation continues may constitute a separate violation. The Administrator may assess the civil penalty by administrative order or may bring an action to assess and collect the penalty in the U.S. District Court for the district in which the person from whom the penalty is sought resides or in which such person's principal place of business is located.

The purpose of this Enforcement Response Policy is to ensure that enforcement actions for violations of EPCRA §313 and the PPA are arrived at in a fair, uniform and consistent manner; that the enforcement response is appropriate for the violation committed; and that persons will be deterred from committing EPCRA §313 violations and the PPA.

For purposes of this document, "EPCRA," "§313" and EPCRA "EPCRA §313" should be understood to include the requirements of the Pollution Prevention Act.

LEVELS OF ACTION

Enforcement alternatives include: (a) no action; (b) notices of noncompliance; (c) civil administrative penalties (d) civil judicial referrals, and (e) criminal action under 18 U.S. Code 1001.

EPA reserves the right to issue a Civil Administrative Penalty for any violation not specifically identified under the Notice of Noncompliance or Administrative Civil Penalty section.

NO ACTION

Revisions to Form R reports

Generally, an enforcement action will not be taken regarding voluntary changes to correctly reported data in Form R reports. Changes to Form R reports are: revisions to original reports which reflect only improved or new information and/or improved or new procedures which were not available when the facility was completing its original submission. Facilities submitting revisions should maintain records to document that the information used to calculate the revised estimate is new and was not available at the time the first estimate was made. A facility which submits a revision to a Form R report which does not meet this description of a change or otherwise calls into question the basis for the initial data reported on the original Form R report will be subject to an enforcement action.

Discussion

Each Form R report must provide estimated releases: it is not acceptable to submit Form R reports with no estimate(s) of releases. Such reports will be considered incomplete reports and subject to an enforcement action as described below. An estimate of "zero" is acceptable if "zero" is a reasonable estimate of a facility's releases based on readily available information, i.e., monitoring data or emission estimates.

Every Form R report submitted after July 1 for a chemical not previously submitted is not a revision, but a failure to report in a timely manner.

Facilities considering whether to submit a revision should refer to the September 26, 1991 Federal Register policy notice which explains for what circumstances a facility should submit a revision and the correct format for submitting a revision. Additionally, the notice explains the purpose of EPA's policy of delaying data entry of all revisions received after November 30th of the year the original report was due until after the Toxic Release Inventory (TRI) database can be made available to the public. Revisions submitted after November 30th will be processed and made available to the public in updated versions of the TRI database. The EPA cannot accept and process revisions to the TRI database on a continuing basis without significantly delaying the public availability of the data. Following on the September 26, 1991 Federal Register policy notice, this ERP adopts the November 30th date to determine the gravity of voluntarily disclosed data quality violations.

NOTICES OF NONCOMPLIANCE (NON)

Summary of Circumstances Generally Warranting an NON

- o Form R reports which are incorrectly assembled; for example, failure to include all pages for each Form R or reporting more than one chemical per Form R.
- o Form R reports which contain missing or invalid facility or chemical identification information; for example, the CAS number reported does not match the chemical name reported.
- o Submission of §313 and Pollution Prevention Act data on an invalid form.
- o Incomplete Reporting, i.e., reports which contain blanks where an answer is required.
- o Magnetic media submissions which cannot be processed.
- o The submission of a Form R report with trade secrets without a sanitized version, or the submission of the sanitized version of the Form R report without the trade secret information.
- o Form R reports which are sent to an incorrect address.

  NOTE: An incorrect address is any address other than that of the U.S. EPA Administrator's office, or other than the address listed in the §313 regulation or on the Form R. Form R reports not received by EPA due to an incorrect address and/or packaging are not the responsibility of EPA and are subject to a civil administrative penalty for "failure to report in a timely manner" violation.

NOTE: The Agency reserves the right to assess a Civil Administrative Complaint for certain data quality errors; see page five for a definition of these types of errors. Generally, these are errors which cannot be detected during the data entry process.

Discussion

A Notice of Noncompliance (NON) is the appropriate response for certain errors on Form R reports detected by the Agency. Generally, these are errors which prevent the information on the Form R from being entered into EPA's database. The NON will state that corrections must be made within a specified time (30 days from receipt of the NON). Failure to correct any error for which a NON is issued may be the basis for issuance of a Civil Administrative Complaint.

The decision to issue NONs for the submission of a Form R report with a trade secret claim without a sanitized version, or of the sanitized version without the trade secret information, is being treated the same as a Form R report with errors. This is a violation of EPCRA §313 as well as the trade secret requirements of EPCRA.

CIVIL ADMINISTRATIVE COMPLAINTS

A Civil Administrative Complaint will be the appropriate response for: failure to report in a timely manner; data quality errors; failure to respond to a NON; repeated violations; failure to supply notification and incomplete or inaccurate supplier notification; and failure to maintain records and failure to maintain records according to the standard in the regulation.

Definitions:

Failure to Report in a Timely Manner This violation includes the failure to report in a timely manner to either EPA or to the state for each chemical on the list. There are two distinct categories for this violation. A circumstance level one penalty will be assessed against a category I violation. A "per day" formula is used to determine category II penalties; see this per day formula on page 13.

- Category I: Form R reports that are submitted one year or more after the July 1 due date.

- Category II: Form R reports that are submitted after the July 1 due date but before July 1 of the following year.

EPCRA §313 Subpart (a) requires Form R reports to be submitted annually on or before July 1 and to contain data estimating releases during the preceding calendar year. Facilities which submit Form R reports after the July 1 deadline have failed to comply with this annual reporting requirement and have defeated the purpose of EPCRA §313, which is to make this toxic release data available to states and the public annually and in a timely manner.

Data Quality Errors: Data Quality Errors are errors which cause erroneous data to be submitted to EPA and states. Generally, these are errors which are not readily detected during EPA's data entry process.[1] Below are the range of actions which constitute data quality errors; generally, these are a result of a failure to comply with the explicit requirements of EPCRA §313:

- o Failure to calculate or provide reasonable estimates of releases or off-site transfers.
- o Failure to identify all appropriate categories of chemical use, resulting in error(s) in estimates of release or off-site transfers.
- o Failure to identify for each wastestream the waste treatment or disposal methods employed, and an estimate of the treatment efficiency typically achieved by such methods, for that wastestream.
- o Failure to use all readily available information necessary to calculate as accurately as possible, releases or off-site transfers.
- o Failure to provide the annual quantity of the toxic chemical which entered each environmental medium.
- o Failure to provide the annual quantity of the toxic chemical transferred off-site.
- o Failure to provide information required by §6607 of the Pollution Prevention Act of 1990 and by any regulations promulgated under §6607 of the Pollution Prevention Act of 1990.

---

[1]EPA's program office may issue Notices of Technical Error (NOTEs) for certain data quality errors which are detected during the data entry process.

- Under the requirements of §6607 of the Pollution Prevention Act of 1990, claiming past or current year source reduction or recycling activities which are not in fact implemented by the facility. This does not apply to activities which the facility may estimate for future years.

- A facility's Form R reporting demonstrates a pattern of similar errors or omissions as manifested by the issuance by EPA of NONs for two or more reporting years for the same or similar errors or omissions.

NOTE: If an error is made in determining a facility's toxic chemical threshold which results in the facility erroneously concluding that a Form R report for that chemical is not required, this is not a data quality error, but a "failure to report in a timely manner" violation.

<u>Failure to respond to an NON</u> When a facility receives a Notice of Noncompliance (NON) and fails to comply with the Notice of Noncompliance, i.e, fails to correct the information EPA requests to be corrected in the NON by the time period specified in the NON, the violation is "failure to respond to an NON." Included here is the failure to also provide the state with corrected information requested in the NON within 30 days of receiving the NON.

<u>Repeated violation</u> This category of violation <u>only</u> applies to violations which would generally warrant an NON for the first time. A repeated violation is any subsequent violation which is identical or very similar to a prior violation for which an NON was issued. Separate penalty calculation procedures (discussed on page 16 under "history of prior violations") are to be followed for violations which warrant a civil administrative complaint for the first violation and are repeated.

<u>Failure to Supply Notification</u> Under 40 CFR §372.45, certain facilities which sell or otherwise distribute mixtures or trade name products containing §313 chemicals are required to supply notification to (i) facilities described in §372.22, or (ii) to persons who in turn may sell or otherwise distribute such mixtures or products to a facility described in §372.22(b) in accordance with paragraph §372.45(b). Failure to comply with 40 CFR §372.45, in whole or in part, constitutes a violation. A violation will be "failure to supply notification" or "incomplete or inaccurate supplier notification."

<u>Failure to Maintain Records</u> Under 40 CFR §372.10, each person subject to the reporting requirements of 40 CFR §372.30 must

retain records documenting and supporting the information submitted on each Form R report. Additionally, under 40 CFR

§372.10, each person subject to the supplier notification requirements of 40 CFR §372.45 must retain certain records documenting and supporting the determination of each required notice under that same section. These records must be kept for three years from the date of the submission of a report under 40 CFR §372.30 or the date of notification under 40 CFR §372.45. The records must be maintained at the facility to which the report applies or at the facility supplying notification. Failure to comply with 40 CFR Part 372.10, in whole or in part, constitutes a violation. Violations will be a "failure to maintain records as prescribed at 40 CFR Part 372.10 (a) or (b)", or a "failure to maintain complete records as prescribed at 40 CFR Part 372.10 (a) or (b)" or "failure to maintain complete records at the facility as prescribed at 40 CFR Part 372.10(c)."

CIVIL JUDICIAL REFERRALS

In exceptional circumstances, EPA, under EPCRA §325(c), may refer civil cases to the United States Department of Justice for assessment and/or collection of the penalty in the appropriate U.S. District Court. U.S. EPA also may include EPCRA counts in civil complaints charging Respondents with violations of other environmental statutes.

CRIMINAL SANCTIONS

EPCRA does not provide for criminal sanctions for violations of §313. However, 18 U.S.C. §1001 makes it a criminal offense to falsify information submitted to the U.S. Government. This would specifically apply to, but not be limited to, EPCRA §313 records maintained by a facility that were intentionally generated with incorrect or misleading information. In addition, the knowing failure to file an EPCRA §313 report may be prosecuted as a concealment prohibited by 18 U.S.C. §1001.

ASSESSING A CIVIL ADMINISTRATIVE PENALTY

SUMMARY OF THE PENALTY POLICY MATRIX

This policy implements a system for determining penalties in civil administrative actions brought pursuant to §313 of the Emergency Planning and Community Right-to-Know Act (EPCRA). Penalties are determined in two stages: (1) determination of a "gravity-based penalty," and (2) adjustments to the gravity-based penalty.

To determine the gravity-based penalty, the following factors affecting a violation's gravity are considered:

- o the "circumstances" of the violation
- o the "extent" of the violation

The circumstance levels of the matrix take into account the seriousness of the violation as it relates to the accuracy and availability of the information to the community, to states, and to the federal government. Circumstance levels are described on pages 11-13.

The extent level of a violation is based on the quantity of each EPCRA §313 chemical manufactured, processed, or otherwise used by the facility; the size of the facility based on a combination of the number of employees at the violating facility; and the gross sales of the violating facility's total corporate entity. The Agency will use the number of employees and the gross sales at the time the civil administrative complaint is issued in determining the extent level of a violation.

To determine the gravity-based penalty, determine both the circumstance level and the extent level. These factors are incorporated into a matrix which establishes the appropriate gravity-based penalty amount. The penalty is determined by calculating the penalty for each violation on a per-chemical, per-facility, per-year basis (see special circumstances for per day penalties on page 13).

Once the gravity-based penalty has been determined, upward or downward adjustments to the proposed penalty amount may be made in consideration of the following factors:

- o Voluntary Disclosure
- o History of prior violation(s)
- o Delisted chemicals
- o Attitude
- o Other Factors as Justice May Require
- o Supplemental Environmental Projects
- o Ability to Pay

The first three of these adjustments may be made prior to issuing the civil complaint.

EXTENT LEVELS

In the table below, the total corporate entity refers to all sites taken together owned or controlled by the domestic or foreign parent company. EPA Regions have discretion to use those figures for number of employees and total corporate sales which are readily available. If no information is available, Regions may assume the higher level and adjust if the facility can produce documentation demonstrating they belong in a lower extent level.

Facilities which manufacture, process or otherwise use **ten times or more** the threshold of the §313 chemical involved in the violation and meet the total corporate entity sales and number of employees criteria below:

	LEVEL
$10 million or more in total corporate entity sales and 50 employees or more.	A
$10 million or more in total corporate entity sales and less than 50 employees.	B
Less than $10 million in total corporate entity sales and 50 employees or more.	B
Less than $10 million in total corporate entity sales and less than 50 employees.	B

Facilities which manufacture, process or otherwise **use less than ten times** the threshold of the §313 chemical involved in the violation and meet the total corporate entity sales and number of employee criteria below:

	LEVEL
$10 million or more in total corporate entity sales and 50 employees or more.	B
$10 million or more in total corporate entity sales and less than 50 employees.	C
Less than $10 million in total corporate entity sales and 50 employees or more.	C
Less than $10 million in total corporate entity sales and less than 50 employees.	C

Discussion

EPA believes that using the amount of §313 chemical involved in the violation as the primary factor in determining the extent level underscores the overall intent and goal of EPCRA §313 to make available to the public on an annual basis a reasonable estimate of the toxic chemical substances emitted into their communities from these regulated sources. A necessary component of making useful data available to the public is the supplier notification requirement of §313, as a significant amount of toxic chemicals are distributed in mixtures and trade name products. An additional goal of §313 is to ensure that purchasers of §313 chemicals are informed of their potential §313 reporting requirements. The extent levels underscore this goal as well.

The size of business is used as a second factor in determining the appropriate extent level to reflect the fact that the deterrent effect of a smaller penalty upon a small company is likely to be equal to that of a larger penalty upon a large company. Ten times the threshold for distinguishing between extent levels was chosen because it represents a significant amount of chemical substance. Thus, the two factors, the amount of §313 chemical involved and the size of business, are combined and used to determine the extent level table.

PENALTY MATRIX

PENALTY MATRIX			
	EXTENT LEVELS		
CIRCUMSTANCE LEVELS	A	B	C
1	$25,000	$17,000	$5,000
2	$20,000	$13,000	$3,000
3	$15,000	$10,000	$1,500
4	$10,000	$ 6,000	$1,000
5	$ 5,000	$ 3,000	$ 500
6	$ 2,000	$ 1,300	$ 200

CIRCUMSTANCE LEVELS

A penalty is to be assessed for each §313 chemical for each facility. There are two "per day" penalty assessments; see page 12 and 13 for further clarification.

The date used to determine the circumstance level for "failure to report in a timely manner" is the postmark date of the Form R submission(s).

All violations are "one day" violations unless otherwise noted.

LEVEL 1

Failure to report in a timely manner, Category I.

LEVEL 2

Failure to maintain records as prescribed at 40 CFR §372.10(a) or (b).

Failure to supply notification; per chemical, per year.

LEVEL 3

Data Quality Errors.

Repeated NON violations.

LEVEL 4

Failure to report in a timely manner, Category II: Per Day formula applies.

Failure to maintain complete records as prescribed at 40 CFR §372.10(a) or (b).

LEVEL 5

Failure to Respond to an NON.

Data Quality Errors which are voluntarily disclosed after November 30th of the year the original report was due.

Incomplete or inaccurate supplier notification; per chemical, per year.

LEVEL 6

Data Quality Errors which are voluntarily disclosed on or before November 30th of the year the original report was due.

Revisions which are voluntarily submitted to EPA but are not reported to the State within 30 days of the date the revision is submitted to EPA.

Failure to maintain records at the facility (40 CFR §372.10(c)).

MULTIPLE VIOLATIONS

Separate penalties are to be calculated for each chemical for each facility. If a company has three facilities and fails to report before July 1 of the year following the year the report was due, a penalty is to be assessed for each facility and for each chemical. Assuming the annual sales of the corporate entity exceed $10 million dollars, the facility has more than 50 employees, and each facility exceeds the threshold limits by more than ten times, the penalty would be $25,000 X 3 or $75,000. If each facility manufactured two chemicals, again at more than ten times the threshold, the penalty would be $25,000 X 3 X 2 or $150,000.

If there is more than one violation for the same facility involving the same chemical, the penalties are cumulative. For example, if a firm reports more than one year after the report was due, and the form also contains errors which the firm refused to correct after receiving an NON, the penalty is $25,000 plus $15,000. However, since it is the same form involved, and since the statute imposes a maximum of $25,000 per violation for each day the violation continues, the penalty which will be assessed should be the one day $25,000 maximum.

PER DAY PENALTIES

Generally, penalties of up to $25,000 **per day** may be assessed if a facility within the corporate entity has received a Civil Administrative Complaint, which has been resolved, for failing to report under §313 for any two previous reporting periods. A Civil Administrative Complaint is resolved by a payment, a Consent Agreement and Final Order, or a Court Order.

Penalties of up to $25,000 **per day** may also be used for those facilities which refuse to submit reports or corrected information within thirty days after a Civil Administrative Complaint is resolved. Such refusal may be the basis for issuing a new Civil Administrative Complaint to address the days of continuing noncompliance after the initial Civil Administrative Complaint is resolved. For example, a respondent may respond to a Civil Administrative Complaint by paying the full penalty, yet not correct the violation; in such a situation, a new Civil Administrative Complaint should be issued.

PER DAY FORMULA FOR FAILURE TO REPORT IN A TIMELY MANNER

The following per day penalty calculation formula is to be used only for violations involving failure to report on or before July 1 of the year the report is due and before July 1 of the following year:

Level 4 Penalty +

$$\frac{(\#\text{ of days late} - 1)\times(\text{Level 1} - \text{Level 4 Penalty})}{365}$$

For example, the penalty for a facility which submitted one Form R report on October 11 of the year the report was due, and met the criteria for extent level A, would be calculated as follows:

$$\$10{,}000 + \frac{(102-1)(\$15{,}000)}{365} = \$10{,}000 + \$4151 = \$14{,}151.$$

CAPS ON PENALTIES

While there is a $25,000 per day per violation maximum penalty under EPCRA §326, which outlines EPA's enforcement authority for EPCRA §313, there are no caps on the total penalty amount a facility may be liable for under EPCRA §313.

ADJUSTMENT FACTORS

The Agency intends to pursue a policy of strict liability in penalizing a violation, therefore, no reduction is allowed for culpability. Lack of knowledge does not reduce culpability since the Agency has no intention of encouraging ignorance of EPCRA and its requirements and because the statute only requires facilities to report information which is readily available. In fact, if a violation is knowing or willful, the Agency reserves the right to assess per day penalties, or take other enforcement action as appropriate. In some cases, the Agency may determine that the violation should be referred to the Office of Criminal Enforcement.

Voluntary Disclosure

To be eligible for any voluntary disclosure reductions, a facility must: submit a signed and written statement of voluntary disclosure to EPA and submit complete and signed report(s) to their state and EPA's TRI Reporting Center within 30 days, or submit complete and signed Form R report(s) immediately to their state and EPA's TRI Reporting Center as indicated on the Form R. In the case of supplier notification violations, the facility must submit a signed and written statement of voluntary disclosure to EPA.

The Agency will not consider a facility to be eligible for any voluntary disclosure reductions if the company has been notified of a scheduled inspection or the inspection has begun, or the facility has otherwise been contacted by U.S. EPA for the purpose of determining compliance with EPCRA §313.

This enforcement response policy establishes two reductions in penalties for voluntary disclosure of violations; the first reduction is a fixed 25%; the second reduction is capped at 25% and can be applied in full or in part according to the extent to which the facility meets the criteria for the second 25% reduction. All facilities which voluntarily disclose violations of §313 (except those identified below) are eligible for the first fixed 25%. The voluntary disclosure reductions apply to the following violations: failure to report in a timely manner, category I and II; and failure to supply notification.

In order to obtain the second reduction for voluntary disclosure a facility must meet the following criteria and explain and certify in writing how the facility meets these criteria:

- o The violation was immediately disclosed within 30 days of discovery by the facility.

- o The facility has undertaken concrete actions to ensure that the facility will be in compliance with EPCRA §313 in the future. Such steps may include but are not limited to: creating an environmental compliance position and hiring an individual for that position; changing the job description of an existing position to include managing EPCRA compliance requirements; and contracting with an environmental compliance consulting firm.

- o For supplier notification violations, the facility provides complete and accurate supplier notification to each facility or person described in §372.45(a) within 60 days of notifying EPA of the violation.

- o The facility does not have a "history of violation" (see below) for EPCRA §313 for the two reporting years preceding the calendar year in which the violation is disclosed to EPA.

This policy is designed to distinguish between those facilities which make an immediate attempt to comply with §313 as soon as noncompliance with §313 is discovered and those which do not.

This enforcement response policy does not allow for voluntary disclosure adjustments in penalties for the following violations because these violations will, in almost all circumstances, be discovered by EPA: failure to maintain records, failure to maintain records according to the standard in the regulation, failure to submit Form R reports containing error corrections or revisions to the state, and failure to supply notification according to the standard in the regulation. In the rare case that a facility identifies such violations and voluntarily discloses them, EPA Regional offices have discretion to adjust the penalty under the "as justice may require" reduction. Consideration of voluntary disclosure for data quality errors is already structured into the circumstance levels: voluntarily disclosed data quality errors are assessed two and three levels lower than data quality errors which are discovered by EPA. Therefore no further "voluntary" reduction is allowed.

NOTE: Reductions available for attitude and for voluntary disclosure are mutually exclusive, as both recognize the facility's concern with, and actions taken toward, timely compliance. Therefore, a facility cannot qualify for reductions in both of these categories.

History of Prior Violations

The penalty matrix is intended to apply to "first offenders." Where a violator has demonstrated a history of violating any section(s) of EPCRA, the penalty should be adjusted upward according to section (d) below prior to issuing the Administrative Civil Complaint. The need for such an upward adjustment derives from the violator not having been sufficiently motivated to comply by the penalty assessed for the previous violation, either because of certain factors consciously analyzed by the firm, or because of negligence. Another reason for penalizing repeat violators more severely than "first offenders" is the increased enforcement resources that are spent on the same violator.

The Agency's policy is to interpret "prior such violations" as referring to prior violations of any provision of the Emergency Planning and Community Right-to-Know Act (1986). The following rules apply in evaluating history of prior such violations:

(a) In order to constitute a prior violation, the prior violation must have resulted in a final order, either as a result of an uncontested complaint, or as a result of a contested complaint which is finally resolved against the violator, except as discussed below at section (d). A consent agreement and final order/consent order (CAFO/CACO), or receipt of payment in response to a administrative civil complaint, are both considered to be the final resolution of the complaint against the violator. Therefore, either a CAFO/CACO, or receipt of payment made to the U.S. Treasury, can be used as evidence constituting a prior violation, regardless of whether or not a respondent admits to the violation.

(b) To be considered a "prior such violation," the violation must have occurred within five years of the present violation. Generally, the date used for the present violation will be one day after July 1 of the year the Form R report was due for failure to report, data quality errors, recordkeeping violations, and supplier notification violations. For other violations, the date of the present violation will be the date the facility was required to come into compliance; for example, for a "failure to respond" violation, the date of the present violation will be the last day of the 30 day period the facility had to respond to a Notice of Noncompliance. This five-year period begins when the prior violation becomes a final order. Beyond five years, the prior violative conduct becomes too distant to require compounding of the penalty for the present violation.

(c) Generally, companies with multiple establishments are considered as one when determining history. Thus, if a facility is part of a company for which another facility within the company has a "prior such violation," then each facility within the company is considered to have a "prior violation." However, two companies held by the same parent corporation do not necessarily affect each other's history if they are in substantially different lines of business, and they are substantially independent of one another in their management, and in the functioning of their Boards of Directors. In the case of wholly- or partly-owned subsidiaries, the violation history of a parent corporation shall apply to its subsidiaries and that of the subsidiaries to the parent corporation.

(d) For one prior violation, the penalty should be adjusted upward by 25%. If two prior violations have occurred, the penalty should be adjusted upward by 50%. If three or more prior violations have occurred, the penalty should be adjusted upward by 100%.

(e) A "prior violation" refers collectively to all the violations which may have been described in one prior Administrative Civil Complaint or CAFO. Thus, "prior violation" refers to an episode of prior violation, not every violation that may have been contained in the first Civil Administrative Complaint or CAFO/CACO.

Delisted Chemicals

For delisted chemicals, an immediate and fixed reduction of 25% can be justified in all cases according the following policy:

If the Agency has delisted a chemical by a final Federal Register Notice, the Agency may settle cases involving the delisted chemical under terms which provide for a 25% reduction of the initial penalty calculated for any Section 313 violation involving that chemical. The reduction would only apply to chemicals delisted before or during the pendency of the enforcement action. This reduction may be made before issuing the Administrative Civil Complaint. Facilities will not be allowed to delay settling Administrative Civil Complaints in order to determine whether the violative chemical will be delisted.

Attitude

This adjustment has two components: (1) cooperation and (2) compliance. An adjustment of up to 15% can be made for each component:

(1) Under the first component, the Agency may reduce the gravity-based penalty based on the cooperation extended to EPA throughout the compliance evaluation/enforcement process or the lack thereof. Factors such as degree of cooperation and preparedness during the inspection, allowing access to records, responsiveness and expeditious provision of supporting documentation requested by EPA during or after the inspection, and cooperation and preparedness during the settlement process.

(2) Under the second component, the Agency may reduce the gravity-based penalty in consideration of the facility's good faith efforts to comply with EPCRA, and the speed and completeness with which it comes into compliance.

NOTE: See note on page 16 regarding the mutual exclusion of reductions for attitude reduction and voluntary disclosure.

Other Factors as Justice May Require

In addition to the factors outlined above, the Agency will consider other issues that might arise, on a case-by-case basis, and at Regional discretion, which should be considered in assessing penalties. Those factors which are relevant to EPCRA §313 violations include but are not limited to: new ownership for history of prior violations, "significant-minor" borderline violations, and lack of control over the violation. For example, occasionally a violation, while of significant extent, will be so close to the borderline separating minor and significant violations or so close to the borderline separating noncompliance from compliance, that the penalty may seem disproportionately high. In these situations, an additional reduction of up to 25% off the gravity-based penalty may be allowed. Use of this reduction is expected to be rare and the circumstances justifying its use must be thoroughly documented in the case file.

Settlement With Conditions (SWC)

Supplemental Environmental Projects (SEPs):

Circumstances may arise where a violator will offer to make expenditures for environmentally beneficial purposes above and beyond those required by law in lieu of paying the full penalty. The Agency, in penalty actions in the U.S. District Courts under the Clean Air Act and Clean Water Acts, and in administrative penalty actions under the Toxic Substances Control Act, has determined that crediting such expenditures is consistent with the purpose of civil penalty assessment. Although civil penalties under EPCRA §313 are administratively assessed, the same rationale applies. This adjustment, which constitutes a credit against the actual penalty amount, will normally be discussed only in the course of settlement negotiations.

Other Settlements With Conditions may be considered by EPA Regional Offices as appropriate.

Before the proposed credit amounts can be incorporated into a settlement, the complainant must assure himself/herself that the company has met the conditions as set forth in current or other program specific policy guidance. The settlement agreement incorporating a penalty adjustment for an SEP or any other SWC should make clear what the actual penalty assessment is, after which the terms of the reduction should be clearly spelled out in detail in the CAFO/CACO. A cash penalty must always be collected from the violator regardless of the SEPs or SWCs undertaken by the company. Finally, in accordance with Agency-wide settlement policy guidelines, the final penalty assessment contained in the CACO/CAFO must not be less than the economic benefit gained by the violator from noncompliance.

Ability to Pay

Normally, EPA will not seek a civil penalty that exceeds the violator's ability to pay. The Agency will assume that the respondent has the ability to pay at the time the complaint is issued if information concerning the alleged violator's ability to pay is not readily available. Any alleged violator can raise the issue of its ability to pay in its answer to the civil complaint, or during the course of settlement negotiations.

If an alleged violator raises the inability to pay as a defense in its answer, or in the course of settlement negotiations, it shall present sufficient documentation to permit the Agency to establish such inability. Appropriate documents will include the following, as the Agency may request, and will be presented in the form used by the respondent in its ordinary course of business:

1. Tax returns
2. Balance sheets
3. Income statements
4. Statements of changes in financial position
5. Statements of operations
6. Retained earnings statements
7. Loan applications, financing and security agreements
8. Annual and quarterly reports to shareholders and the SEC, including 10 K reports
9. Business services reports, such as Compusat, Dun and Bradstreet, or Value Line.
10. Executive salaries, bonuses, and benefits packages.

Such records are to be provided to the Agency at the respondent's expense and must conform to generally recognized accounting procedures. The Agency reserves the right to request, obtain, and review all underlying and supporting financial documents that form the basis of these records to verify their accuracy. If the alleged violator fails to provide the necessary information, and the information is not readily available from other sources, then the violator will be presumed to be able to pay.

SETTLEMENT

Any reductions in penalties are to be made in accordance with this penalty policy. In preparing Consent Agreements, Regions must require a statement signed by the company which certifies that it has complied with all EPCRA requirements, and specifically §313 requirements, at all facilities under their control.

Any violations reported by the company or facility in the context of settlement are to be treated as self-confessed violations or treated as a failure to report in a timely manner if the company has not submitted the report. If a Region wishes to enter into a Settlement Agreement for the facility/company to audit its facility/company, then the Consent Agreement and Final Order may contain this agreement. A Region may choose to agree to assess prior stipulated penalties for the violations found during the compliance audit, or may choose to assess any such violations in accordance with this enforcement policy. Reductions for compliance audits cannot exceed the after-tax value of the compliance audit. Finally, as stated above, a cash penalty must always be collected from the violator regardless of the SEPs or SWCs undertaken by the company.

AMENDMENT for 1991 Reporting Year Only

Due to the unusual circumstances in finalizing and distributing the revised Form R for use beginning with calendar year 1991 reports (reports due on July 1, 1992), the following amendment to the Enforcement Response Policy is issued:

Penalty Assessment for Failure to Report in a Timely Manner

One element of the Per Day Penalty Formula on page 14 is the number of days late a facility submits its Form R reports. For the 1991 reporting year only, the number of days late will be calculated beginning on September 2, 1992. Thus, if a facility submits its Form R report on September 15, 1992, the number of days late should be calculated as 14.